D0220589

Alice Arndt

Seasoning Savvy
How to Cook with Herbs, Spices, and Other Flavorings

More pre-publication
REVIEWS, COMMENTARIES, EVALUATIONS . . .

"**T**his book is chock-full of little-known bits of wisdom, lore, and practical information. It shows us how to select and store herbs and spices, and most important, how to get the most flavor from seasonings when cooking. *Seasoning Savvy* will help you discover the wonderful flavors that seasonings can bring to a wide range of foods. I know that I will be making good use of this book."

Carole Bloom, CCP
Author, *All About Chocolate: The Ultimate Resource for the World's Favorite Food, Sugar & Spice,* and *Truffles, Candies, & Confections*

"**A**lice Arndt's passion for the flavors that spice up our lives is felt on every page of this indispensable reference work. Equally at home in the tropical realm of exotic spices and the cozy kitchen garden, Arndt draws the reader into an exciting world of flavors. This book belongs on every library shelf, but it should also be brought into the kitchen. Arndt provides a comprehensive list of seasonings and offers reassuring advice on their culinary uses, so that even novice cooks can use flavorings like asafetida with aplomb. She explains how to dry herbs in the refrigerator so that they don't get dusty, and how to shine copper bowls with lemon and salt to avoid toxic cleansers.

In addition to offering detailed descriptions of the seasonings that define different cultures, this book identifies the world's most important spice mixtures and gives recipes for recreating them. The index is essential for anyone who has ever wondered about the difference between nigella, onion seed, black sesame, black cumin, and Russian caraway seed. You'll find this answer, and many more, in Alice Arndt's wonderful book."

Darra Goldstein, PhD
Professor of Russian, Williams College; Author, *The Vegetarian Hearth; A Taste of Russia; The Georgian Feast*

"**A**lice Arndt's *Seasoning Savvy* is a well-written and wonderfully comprehensive exploration of the world of herbs, spices, and aromatics—at once authoritative and easy to use. There is inspiration and good advice for everyone from home cooks and restaurant chefs to culinary historians and food scholars."

Nancy Harmon Jenkins
Author, *The Mediterranean Diet Cookbook*

The Haworth Herbal Press
An Imprint of The Haworth Press, Inc.

Seasoning Savvy
*How to Cook with Herbs, Spices,
and Other Flavorings*

Seasoning Savvy
How to Cook with Herbs, Spices, and Other Flavorings

Alice Arndt

The Haworth Herbal Press
An Imprint of The Haworth Press, Inc.
New York • London • Oxford

Published by

The Haworth Herbal Press,® an imprint of The Haworth Press, Inc., 10 Alice Street, Binghamton, NY 13904-1580

"The Prayer of the Bee," from *Prayers from the Ark* by Carmen Bernos De Gasztold, translated by Rumer Godden, translation copyright © 1962, renewed 1990 by Rumer Godden. Original copyright 1947, © 1955 by Editions du Cloître. Used by permission of Viking Penguin, a division of Penguin Putnam Inc.

Cover design by Monica L. Seifert.

In-text photographs by Alice Arndt.

Library of Congress Cataloging-in-Publication Data

Arndt, Alice.
    Seasoning savvy : how to cook with herbs, spices, and other flavorings / Alice Arndt.
        p.   cm.
    Includes index.
    ISBN 1-56022-031-7 (hc: alk. paper). 1-56022-032-5 (pbk: alk paper).
    1. Cookery (Spices) 2. Spices. 3. Cookery (Herbs) 4. Herbs. I. Title.
TX819.A1 A764   1999
641.6'383—dc21
                                                                          99-20217
                                                                              CIP

ABOUT THE AUTHOR

Alice Arndt, is a food historian and cooking teacher who teaches, writes, and lectures about food, its history, and the spices used in cooking. Her experiences living and traveling in many parts of the world have brought her a wealth of recipes and information on these subjects, which she shares enthusiastically with her audiences. Ms. Arndt has lectured and demonstrated at the Smithsonian Institution, Harvard's Semitic Museum, the Archaeological Institute of America-Houston Society, the Houston Public Library and Houston's International Festival, the Oxford Symposium on Food and Cookery, the World History Association—Texas, and for several Culinary Historian groups. Her lectures, interviews, and cooking classes have also aired on radio and television. Ms. Arndt has written for numerous magazines and newspapers, such as *Americana, AramcoWorld, The Christian Science Monitor, Food History News, Horticulture, The Houston Chronicle,* and *The Radcliffe Culinary Times.* She contributed to *The All New Joy of Cooking* published in 1997, as well as to the book *Women Travel* (1990) and the *Arab World Studies Notebook* (1998). She is now editing *The Dictionary of Culinary Biography.*

CONTENTS

Acknowledgments

It gives me great pleasure to acknowledge and thank the many people who have helped me with this book and with my study of seasonings in general. There is not enough room here to list all the specifics of their generous sharing of expertise, resources, and ideas. Without their interest and encouragement, this book would have taken me even longer than it did to write. My sincere thanks to one and all!

Hicham Aboukinane; American Spice Trade Association; Anne Marie Weiss Armush, *Arabian Cuisine*; Miss Willie Atkins; Cathleen Baird, Hilton College Archivist, University of Houston; Najmieh Batmanglij, *Food of Life*; Michael and Lucia Bettler, Lucia's Garden; Botanical Research Institute of Texas, Ft. Worth; Professor George E. Brooks, Indiana University; Kari Wulff Çağatay; Terrie Chrones; the late Sophie D. Coe, *America's First Cuisines*; Shirley Corriher, *CookWise*; Kathleen Curtin; Plimoth Plantation; LaVerl Daily; Le Panier Cooking School; Dat'l Do It, Inc.; June di Schino; Leonor DuBose, Chata's Mexican Cuisine; Fuchsia Dunlop; Stephen Facciola, *Cornucopia: A Source Book of Edible Plants*; Virginia Ferguson, Arthur D. Little, Inc.; Dr. K. N. Gandhi, Gray Herbarium Index Bibliographer, Harvard University; Sandy Garson; Dalia Carmel Goldstein; Professor Darra Goldstein, Williams College; Barbara Hauke; the many talented and efficient people I worked with at The Haworth Press; Madalene Hill and Gwen Barclay, Flavor Connection; Filiz Hösükoğlu; Dorothy Huang, *Chinese Cooking*.

Special thanks to the Indian Spices Board and the Cardamom Board, particularly N. Bharathan Pillai, S. Dutt Majumder, and Mrs. Usha; also in Kerala, Eapen George, Samuel Eddy, and Mr. and Mrs. K. M. Kutappa of Arnakal Estate.

Eve Jocknowitz; Benjamin H. Kaestner III, McCormick & Co.; Lynn Kay; Patricia M. Kelly; Dr. Sandy Ketelson; the library staff,

Royal Botanic Gardens, Kew; Lewis & Neale, Inc.; Lisy Corporation; Suzanne Lombardi, Dancing Deer Baking Company; Glenn Mack and Asele Surina, Creative Cuisine and Catering of Austin, Texas; Paula Marcoux, Plimoth Plantation; Rick McDermott; Melissa's; Thomas Miller, McCormick & Co.; Mounirah Mosly; Mr. Moy, San Antonio Botanical Garden, Texas; May Mansoor Munn; Hanadi Neemani; Gwen Nguyen, Taste of Asia; Sandra L. Oliver, *Food History News*; Angelica Orozco; Dr. Marcia Pelchat, Monell Chemical Senses Center; William Penzey, Sr., Penzey's Spices; Judith Pierce Rosenberg.

I am deeply indebted to the Schlesinger Library, Harvard University; to Barbara Haber, Curator of Printed Books; Barbara Ketcham Wheaton, Honorary Curator of the Culinary Collection; Wendy Thomas, Public Service Librarian, and all the staff; I would also like to express my appreciation for the many generous donors who have contributed useful books to the culinary collection.

Lucy Seligman, *Gochiso Sama* newsletter; Margaret Shaida, *The Legendary Cuisine of Persia*; Professor Beryl Simpson, University of Texas at Austin; Dr. Behnaz Smalley; Nancy Stutzman, University of Massachusetts Extension; Muhammad A. Tahlawi; Martha Taylor; Betty Teel; Agni Thurner; Dr. Joyce Toomre; Chef Eddy Van Damme, Houston Community College Culinary Arts Program; my mentor in Indian cooking, Suneeta Vaswani, Suneeta Vaswani's Indian Cuisine; Norma Jost Voth, *Mennonite Foods & Folkways from South Russia*; Heidemarie Vukovic, Chocolate Harmony.

The death of F. H. "Ted" Waskey, PhD, Certified Executive Chef and Registered Dietitian, was a loss both to this book and to the food world in general.

Barbara Ketcham Wheaton, *Savoring the Past*; Ken and Amber Wickwire; Ann Wilder, Vanns Spices Ltd.; Paula Wolfert, *Couscous and Other Good Food from Morocco*; Susan Wood; Ann Woodward, *The Exile Way*; Nan Wright, Houston Baptist University; Dr. Charles J. Wysocki, Monell Chemical Senses Center.

Finally, but without end, I am grateful to my encouraging and helpful family—my children Nicole Arndt and Elizabeth and Eric Tosaris, and my husband Robert Arndt.

Introduction

So much can be said about herbs and spices! These plants are important for their medicinal and nutritional properties, as well as for more mythical qualities ascribed to them in folklore. They have great commercial value and also provide simple pleasures to those who grow them at home. Many herbs and spices have also served as cosmetics, dyestuffs, perfumes, or insect repellents. Flaxseed even gave us a floor covering—linoleum! Throughout their long association with human beings, these small vegetable products have had a significant, sometimes tumultuous, effect on the course of history.

But this is a book about the culinary virtues of these useful plants. It is not a cookbook but an adjunct to every cookbook, telling you how to get the most out of the seasonings you use to flavor your food. This book will guide you in the selection of seasonings, in their storage, and finally, in using them to prepare a delicious dish. Substitutes are suggested in case you have difficulty finding the seasoning called for in a recipe. Of course, a substitute is always just an approximation, providing a similar taste or effect; but even though it is not identical, it will still be good.

You are urged to grow a few herbs yourself. Many kinds are not available at all in supermarkets, and what is there cannot be as fresh as a homegrown herb. Having your own garden will also give you access to the herbal flowers, whose use is mentioned in this book. Even with only a few pots of herbs on a sunny windowsill, you can work wonders with the flavors of your cooking.

The two terms *seasonings* and *flavorings* are used in a sweeping fashion to encompass any substance that adds a flavor to food, and thus includes all the herbs and spices in the book, plus **chocolate** and **vanilla** extract as well as some gums and resins. Note that *Seasoning Savvy's* chapter "Individual Seasonings" incorporates such foods as **almonds**, **citrus fruits**, and **coconuts**, because these are frequently used in cooking primarily for the flavor they impart to a dish. Although *seasoning* and *flavoring* are often used inter-changeably, cooks have a tendency to speak of "seasoning" a savory dish, and "flavoring" a sweet one.

Condiment is a similarly ambiguous term, but it is mainly restricted to flavorful concoctions, usually somewhat liquid, such as ketchup, chutney, or Worcestershire sauce, that are served separately and are added at the table at the diner's discretion.

Fundamentally, *Seasoning Savvy* is about flavor, and how to use seasonings to achieve the flavors you like. Thus, there is a discussion in this chapter of what flavor actually is and how we perceive it. The suggestions in the "Culinary Practice" chapter for storage of spices and herbs, toasting, chopping, and so on, are all designed to maximize their flavor—or in any case, to control it: for some seasonings, such as **garlic** and **chiles**, you are also told how to reduce their intensity. Individual seasonings are discussed in detail in the following chapter, which is the heart of the book. And the final chapter deals with established blends and combinations that are sometimes called for in a recipe, from standard spice mixtures to flavored butters, oils, and vinegars.

INCREASING POPULARITY OF SEASONINGS

A phenomenon of American culinary history is the towering rise in the consumption of seasonings between the beginning and the end of the twentieth century. For much of the nineteenth century cookery books advocated restraint and caution in the use of seasonings, and in the first decade after the turn of the century some authors argued vehemently against the use of any spices or condiments at all on the grounds of taste, health, and morality.

As the century wore on, interest in spices and herbs began to grow slowly. Estimates put per capita consumption in 1920 at 0.78 pounds. In 1933, the Herb Society of America was formed, and did much to bring attention, respectability, and popularity to cooking with herbs. A few culinary writers took up the subject, but noted the prevailing misconceptions about the use of herbs and anticipated some resistance to their recipes.

After World War II, consumption of seasonings began to shoot up rapidly and steadily. GIs were credited with bringing home a taste for **oregano**, the "pizza herb," and other foreign flavors, along with a new global awareness. In the 1950s and 1960s, herb and spice cookbooks and guides to seasonings—such as this one—came out on a tide of publishing that has never ebbed. But there remained a certain ambivalence about the use of seasonings. Even pamphlets put out by spice companies found it necessary to caution against overseasoning.

Whatever the cookbooks told them, Americans *were* spicing up their food. In 1971, the American Spice Trade Association calculated that consumption of spices, herbs, and other seasonings had increased more than 40 percent in less than a decade. Per capita consumption in 1972 had risen to 1.41 pounds per year, approximately double the level of a half-century earlier. United States spice imports reached record levels almost annually throughout the 1980s, as professional chefs and home cooks alike responded to the country's increased appreciation for ethnic and regional foods from all over the world.

The rate of increase in the use of seasonings may have begun to level off in the last decade of the century, but absolute amounts continue to rise. In 1990 our average consumption exceeded three pounds per person for the first time, and has not dipped below that mark since. Hot spices, such as black **pepper**, **chiles**, **mustard**, and **ginger** are particularly popular. As the century closes, with travelers and immigrants introducing new foods and flavors, and with the food industry going global, American cuisine is ever more adventurous and inventive—and well-seasoned.

HERB OR SPICE?

What is the difference between an herb and a spice? Many people ask that question, but almost nobody likes the answer. The truth is, the words *herb* and *spice* are not well defined, and every attempt to spell out a distinction between them soon runs into exceptions to the rule. No matter how you describe the two categories of seasoning, it seems you can always find some substance or other that fits into both groups, or neither.

Speaking generally, we know what an herb is. It is first of all a plant native to the temperate regions of the earth. Our word derives from the Latin *herba,* meaning "grass," and most herbs come from soft green leafy plants; unlike shrubs and trees, these do not develop woody stems. Of course, we don't want every temperate-zone non-tree or nonshrub to qualify for the lofty status of herb; some are just ornamentals or, worse, weeds. We know that herbs are aromatic and flavorful, with possible medicinal or even aphrodisiac properties, and several definitions specify the presence of some fragrant essential oil in the leaves that gives them their potency.

Spices, on the other hand, are the precious commodities of tropical regions. The spices are plant parts other than leaves, such as bark, roots, seeds, sap, or flower buds; they are often dried. They, too, have outstanding fragrances and strong tastes, and often serve additionally as dyestuffs, cosmetics, or magic charms.

These distinctions work well enough in general, but many exceptions immediately spring to mind: **Rosemary**, for example, is a fine temperate-zone herb, but in suitable conditions it can form an enormous, woody-stemmed hedge, bristling with spiky little leaves, penetrating in its perfume and buzzing with bees; anything but soft! And is **lavender** an herb? Certainly not tropical—it is decidedly unhappy in too warm a climate—it develops persistent woody stems; and we use both its flowers and its leaves. Or consider the **bay leaf**—plucked from a Mediterranean *tree*!

Many soft temperate-zone plants, such as **anise**, **caraway**, **cumin**, and **fennel**, are valued not for their leaves but for their aromatic small fruits, which we call seeds. And then there are **poppy** seeds, which contain no essential oil at all. Our definitions of *herb* clearly have both gaps and overlaps.

The Herb Society of America has defined an herb as "a plant which may be used for physic, flavor or fragrance," and members declare their dedication to plants "for use and for delight," a very inclusive brief indeed. In the end, it seems impossible to improve upon the whimsical description given twelve centuries ago by Alcuin, master of Charlemagne's Palace School and one of the great educators of the Dark Ages. In a collection of rhetorical riddles for the edification and enjoyment of his students, Alcuin poses the question, "What is an herb?" The answer: "The friend of physicians and the praise of cooks."

Spices, too, resist being stuffed into a lexicographical pigeonhole. Many spice plants even refuse to be confined to the tropics. Conversely, **allspice**, a loyal native of Jamaica and Central America, does not grow well in other locations, tropical or not. Spices in the ginger family, such as **cardamom**, **galangal**, **turmeric**, and **ginger** itself, come from plants that are never woody. And both ginger and galangal are often used—and increasingly available —fresh, rather than in dried form.

Never mind; there is no reason to make too much of this argument. The herb/spice distinction is essentially a holdover from medieval Europe, when an herb was what you could grow in your own herb garden or physic garden, while a spice had to be imported from some distant land. Given the great expense and high prestige of using spices in the Middle Ages, the distinction made a difference. But today, wherever we live, both spices and herbs are imported from around the world for our enjoyment. Our experiences in the herb garden differ also: there are temperate places in the United States, such as East Texas, where we cannot grow **tarragon** in our gardens and we cannot *stop* growing ginger!

In keeping with tradition, the leafy temperate-zone plants like **parsley**, **sage**, **oregano**, and **thyme** will probably always be called "herbs"; **pepper**, **cinnamon**, **cloves**, and **nutmeg** are classically "spices"; and we needn't worry about the ambiguous ones like **saffron** or **lemon grass**. They are simply seasonings, flavorings, useful plants—and it is the aim of this book to make them all "the praise of cooks."

HERB OR 'ERB?

Do you season your food with a "herb" or with an "'erb"? Until the nineteenth century, the word was pronounced "'erb" throughout the English-speaking world, at which time the British began to say "herb" while Americans continued to drop the initial *h*. Today, American dictionaries consider either pronunciation to be correct, but the question can still generate some rather heated debate in local Herb Society meetings! In contemporary usage, there seems to be a slight shift toward pronouncing the *h*—but you can confidently maintain your use of whichever form you are comfortable with.

FLAVOR

Taste and smell, described as our two chemical senses, work together in our perception of the flavor of food. The process is extremely complex, and not completely understood. Neither the tongue nor the nose alone is sufficient to experience flavor; we must have the combination of taste and aroma, as the brain registers signals from the taste buds in the mouth and from the olfactory sensory cells at the roof of the nasal passage. In addition, the sense of touch is inextricably involved in our experience, as we react to the "mouthfeel" of food, determined by its texture and consistency. The temperature of the food, as well as the presence of certain chemicals which we describe as "hot" (as in chiles) or "cool" (as in mints) are also factors. Piquant foods actually cause a degree of pain, in a way that offends some eaters but excites and pleases others. The look of the food before we eat is important, as are the sounds we hear—crunchy, squeaky, slurpy, fizzy, and so on—as we chew and swallow. And finally a feeling of satisfaction, satiety, or even bloatedness, or the discovery that "it tastes like *more*," all contribute to the profound experience of savoring our food. Moreover, these sensations are modified by our mood at the moment, our state of health, our expectations, nostalgia, and the taboos and aversions we have learned.

Our sense of smell, while perhaps dull in comparison to most other mammals', is nonetheless extremely subtle, and we can discriminate among thousands of different scents. (The figure 10,000 has been suggested for the number of scents recognized by humans, although this has not been scientifically established.) In contrast, the gustatory sense—the perceptions communicated by the taste buds—is usually regarded as rather crude, being limited to just four basic tastes: sweet, sour (or acid), bitter, and salty. This theory is by no means universally accepted, however; some physiologists and some culinary cultures suggest that there are one or more additional basic tastes, variously described as earthy, metallic, astringent, alkaline (soapy), or spicy (pungent). It is difficult to specify precisely what is meant by these terms, and some of them may overlap. Our problems in identifying the basic tastes may reflect a physiological fact: Although the question of how many basic tastes there are goes back at least to Aristotle, no one has yet demonstrated that such a thing as a basic taste actually exists.

Finally, our language is very poor in descriptors for taste, and much of the discussion of a taste sensation is often done by analogy with other senses: for example, a flavor may be "bright" or "sharp." Just as wine tasters have developed a special vocabulary to describe various wines, so tasters in spice companies and in synthetic-flavor laboratories have worked to devise a set of terms for taste; so far, however, these have not been standardized and their specific definitions are by no means widely known.

A number of questions remain about how the taste buds actually work. There are three different types of papillae, or bumps, on the surface on the tongue, each one containing one or more taste buds. The number of these tiny protuberances varies widely among individuals and also decreases with age in a given person, but the average mouth contains about 9,000 of them. Their exact location varies as well; they are located mostly on the tongue, but a few may appear on the soft palate, the hard palate, or the pharynx.

Different areas of the tongue differ in sensitivity to the various tastes. Maximum sensitivity to sweetness occurs at the tip of the tongue, and to bitterness at the back. Salty and sour tastes are mainly detected by the taste buds on the sides of the tongue, with saltiness registering predominantly toward the front and sourness in the back. Although this describes only the areas of greatest sensitivity—every taste registers everywhere to some extent—this phenomenon explains why, when you are tasting a dish for seasoning, you should be sure to use your whole mouth, not just the tip of your tongue.

The combination of different tastes affects the perception of the whole. Thus the standard method for correcting a dish that is too salty is to add a little sugar; there is no less salt in the dish afterward, but it tastes less salty. However, the situation is far more complicated than the mere dimming of one taste by another, since salt is known to "bring out" the other ingredients; that is, it increases the perception of other tastes. Monosodium glutamate, called MSG, is regarded as another flavor enhancer, and it has also been suggested as an additional basic taste.

Temperature, too, affects taste perception. A sweet substance will taste sweeter in a warm dish than in one at room temperature, while the opposite is true for bitter foods: they taste more bitter when lukewarm than when they are hot. A dish wants more salt after it has cooled; and all tastes pale at the extremes of hot or cold.

Experiments indicate that our perception of flavor is influenced by our hormonal condition as well—recall the old stories about pregnant women craving dill pickles and ice cream. Both sickness and many of the medicines used to treat it affect our sensitivity to tastes. Our nutritional state is also important: every cook knows that "hunger makes the best sauce," and a craving for a certain flavor sometimes indicates a nutritional need.

Finally, genetic differences in sensitivity to tastes may lead to differences in flavor perception and, consequently, flavor preferences or aversions. Thus, as sages in all ages have recognized, *chacun à son goût* (each to his own taste).

The savvy cook manipulates combinations of various flavors and other factors to create something that is greater than the sum of its parts. A well-made dish is one not easily analyzed, whose various ingredients blend into a unified whole, with one or perhaps two flavors predominating and backed up by a harmonious chorus. (However, the Japanese chef takes exception to this rule. The relatively few flavors employed in each dish are meant to remain apart, for contrast and so that each may be appreciated individually. This explains why Japanese seasonings are frequently added at the table; strong flavors, especially ginger and wasabi, are served as condiments to be added by the eater.)

THE BENEFITS OF FLAVOR

Flavorful food is a joy and a pleasure that needs no excuse or justification. Nevertheless, it is actually healthy as well.

Flavor can guide us in making nutritional choices, as mentioned in the preceding paragraphs, but that is not all. What the physiologists call the hedonistic aspect of taste, in this case the sheer delight that good seasonings can bring to eating, apparently has real health benefits. Good flavor, or even anticipated good flavor suggested by the appetizing appearance and smell of the food, causes saliva to

gather in the mouth. Saliva contains an enzyme that begins digestion, especially of the starches, and it also helps moisten and soften
the food for swallowing. The stomach also responds to pleasant
flavors by secreting extra gastric juices. So the better the flavor, the
better our digestion, and the more comfortable, healthy, and well-
nourished we are!

Consider the so-called "French paradox": the traditional French
diet is heavy with cholesterol-laden foods—cream, butter, eggs, red
meats, cheeses, goose liver—and French eaters do have high cho-
lesterol levels in their blood, but they do not have the high rate of
heart attacks that American doctors associate with this condition.
One of the most convincing hypotheses put forward to explain this
paradox is that in France the traditional relaxed enjoyment of good,
flavorful food at the table is what keeps the heart attacks away.

Seasonings can make appetizing the foods that the doctor orders,
be they low-salt, no-cholesterol, or simply more vegetables. A
spoonful of sugar makes the medicine go down; a handful of herbs
nicely compensates for salt; and intriguing spices make up for low-
ering the fat content.

In seasoning food, respect your own personal tastes and require-
ments. Use your favorite herbs and spices to create dishes to please
yourself. You'll be happier and healthier—and you'll probably find
someone else who utterly adores your cooking.

HOW TO USE THIS BOOK

Herbs and spices are listed in alphabetical order in the "Individual Seasonings" chapter, under the common name most generally
used in the United States. Many plants have acquired several additional names; these are given as well, and are also listed in the
Index. If you cannot find the name you know in the alphabetical
listing of individual seasonings, consult the Index, where it will
probably be entered as an alternative name. Do not take spelling too
seriously: many of our common names come from other languages,
and have been transliterated in different ways.

The plants' scientific names are also given (in italics) to help
ensure proper identification. Note, however, that plant taxonomists
are not always in complete agreement over the status of a name and,

furthermore, that the accepted names of many herbs and spices have changed in the course of the twentieth century, as botanists have gained new knowledge through modern techniques and as plant nomenclature has become increasingly systematized and codified.

In this book, a few herbs and spices are discussed only in conjunction with a related seasoning and do not appear in the Contents. They are, however, included in the Index. To paraphrase, for emphasis: not every plant discussed in this book will be found in the Contents; consult the Index for a complete listing.

Common seasoning blends are listed separately, in alphabetical order, in the "Flavor Combinations" chapter.

Culinary Practice

Individual spices appearing in **bold** type below are discussed in detail in the following chapter, where they are listed in alphabetical order. Seasoning blends in **bold** type are described in the chapter "Flavor Combinations."

SELECTING SEASONINGS

Fresh Herbs

For seasoning satisfaction, try growing a few herbs for yourself. Homegrown plants are fresher and more flavorful, and many more options are available to you—such as **anise** leaves, fresh **angelica** stalks, and the many beautiful edible herbal flowers only rarely seen in the supermarket. With just a little TLC, your favorite herbs will flourish in a pot on a windowsill or in a container on a patio or in a sunny spot in the garden outside, or wherever is appropriate for your situation. When selecting a site, avoid areas near busy roads or parking lots, because your plants can absorb dangerous substances from the exhaust fumes of cars. Never use pesticides in the herb garden. Many herbs repel insects naturally and will extend their protection to other plants located nearby; for example, pennyroyal chases away ants and fleas, **garlic** repels aphids and weevils, tansy deters flies, and **chiles** are unpopular with tomato and cabbage worms.

To select herb plants for your garden—or bunches of cut fresh herbs in the market—"disturb" the leaves with your hand; there should be ample volatile oils to impart an aroma to your fingertips, and the smell should be pleasant. The possibility of making this test is one reason why it is better to grow herbs from seedlings rather than from seeds, unless you know the parent plant.

When cutting herbs in the garden for cooking, avoid tough stems, and be careful not to bruise the leaves; scissors work better than a

knife. It really is best to gather herbs early in the morning, if possible, just after the dew has dried and before the sun begins to
evaporate the volatile essential oils. After picking, put the whole
leaves into an airtight container and keep them in the refrigerator
for the remainder of the day until you are ready to use them. But
don't worry if you have forgotten to gather an herb in the morning:
fresh herbs are good whenever you have the time to cut them, and
they will also reward last-minute use.

Spices

Spices come in many grades, roughly determined by the amount
of essential oil that they contain, but also influenced by other factors
such as size, source, and the success (or failure) of a recent harvest.
Price is a general guide to quality, but not an infallible one. The
larger spice companies sell different grades of spices with distinctive packaging, but your supermarket manager generally selects just
one grade to offer you. Specialty spice importers often take pride in
selling the finest spices available from each harvest. If you ever
open a recently purchased bottle or package of spices and find the
contents dry and odorless, take it back for a refund!

STORAGE

The aim of storing spices and herbs properly is to preserve their
aroma and pungency, and prevent loss of their volatile oils. As the
word "volatile" implies, these essential oils will quickly fly out into
the air, if given the opportunity. That's why whole spices are much
more flavorful than ground spices, not only because grinding releases much of the oils but also because these essences can escape
much more easily when the spice is stored in ground form. Thus, all
seasonings should be kept in airtight containers. For this reason
also, spices and herbs should not be bought in bulk; frequent pur-
chases of small amounts of fresh seasonings (see "Selecting Seasonings" in this chapter) will keep your flavors consistently bright
and clear.

Heat encourages evaporation of the essential oils, so spices
should never be stored near the stove or in other warm parts of the

kitchen. Moisture is also hard on spices and herbs, and many are sensitive to light as well; spice containers need to have tight lids and to be made of dark glass or opaque materials or be kept in a dark cupboard located in a cool part of the kitchen.

So why do we have a profusion of spice racks fitted with clear glass bottles and designed for hanging on the wall right out in the light and heat of the kitchen? That's a very good question! Supposedly, these racks make the seasonings handy for the cook, who simply reaches up, perhaps impulsively, from the stove or kitchen counter for a powder or some flakes to sprinkle into the dish. Whoa! The cook with seasoning savvy assembles all ingredients before beginning to cook, and toasts or chops and measures the spices and herbs in advance. It is not necessary to follow a recipe slavishly, but thinking through which seasonings are likely to enhance the dish, and tasting and adjusting as it progresses, are important aspects of flavoring. None of this implies a need to have a spice rack within reach of the work space, especially not a spice rack full of bleached, bland seasonings!

The best instrument for assessing the freshness of your herbs and spices is your nose. If the aroma is elusive, the flavoring is losing its potency. Many sources tell you to throw out your old spices and replace them at this point, but it is difficult to part with these precious commodities in so cavalier a manner. Properly dried herbs and spices don't spoil; they just become weaker. (Certain oily seeds, however, such as **coconut** and **poppy** seeds, have to be watched for rancidity.) When your herbs and spices are becoming outdated, increase the quantity used in each dish. Increase the amounts gradually, starting with perhaps an additional half of the measure called for, and taste to see if that is sufficient. Add a little more of the seasoning if the dish is still too bland. You will thus get more flavor from your spices and herbs at the end of their useful lives, and simultaneously use them up faster. When they are gone, you can buy fresh spices in small amounts and store them carefully. It is useful to make a note of the purchase date on the spice jar.

The sulfur compounds in **asafetida** are an effective natural insect deterrent and, if you do not find it too offensive, you might want to place a lump of asafetida on a shelf of your spice cupboard to keep insects out of the seasonings.

Chiles

Every kind of capsicum should be kept in a tightly closed container in the refrigerator. This means your **chili powder**, **paprika**, red pepper flakes, or ground cayenne, as well as all whole **chiles** whether fresh, dried, or smoked. Not only does the refrigerator keep their colors bright and their flavor fresh, but it keeps them safe from weevils and other insects that enjoy capsicums as much as we do.

Seeds

True seeds such as canola seeds, flax seeds, **mustard** seeds, **poppy** seeds, or **sesame** seeds typically have a high oil content, and most of them should be stored in the refrigerator to prevent rancidity. Sesame seeds are exceptional, being renowned for their keeping qualities, even in warm climates. Mustard seeds also have a long shelf life, and will easily last a year in a tightly sealed container stored in a cool, dark place.

Note that some seasoning seeds are actually the tiny dry fruits of their plants; they are commonly called seeds because of their size. These include various members of the parsley family: **ajowan**, **anise**, **caraway**, **celery** seeds, **coriander**, **cumin**, **dill** seeds, and **fennel** seeds. These spices do very well when simply stored in air-tight containers in a cool, dark spice cupboard.

DRYING HERBS

Ovens, microwaves, and even direct sunlight are all too hot for drying fresh herbs well; the heat evaporates too much of the plants' essential oils, causing an unacceptable loss of flavor. Proper drying of herbs requires cool, circulating air that is not too humid. Some climates offer these conditions naturally, and it has long been a tradition in Europe and in many of the cooler regions of America to hang bunches of herbs, freshly gathered in late summer, from the rafters until they have dried. Unfortunately, this method often results in rather dusty leaves, especially if they are hung in large bunches, which take longer to dry completely.

Luckily, most houses in America today are provided with an absolutely perfect dust-free, dark herb dryer with cool, regularly circulating dry air: it's your frost-free refrigerator! Simply spread out the sprigs of herbs on a rack or a plate, uncovered, and place them on an open shelf for three or four days until the leaves are dry and crisp. If you use a plate instead of a rack, turn the herbs over each day. When dry, the herbs will still be green and will retain almost all their flavor. The leaves can then be stripped from the stems, sealed tightly in a spice jar, and stored in a cool, dark cupboard with all your other spices and flavorings.

FREEZING HERBS AND HERBAL DISHES

Fresh herbs can be kept in a freezer for as long as six months. They will maintain their color and flavor, but their texture will suffer. The leafier the herb, the less well it will freeze. **Rosemary**, for example, does quite well; **parsley**, on the other hand, turns mushy as soon as it thaws. Although these soft herbs can still be used for flavor in a stew or soup or sauce, their texture and appearance are too unappealing for most other uses. But if your heart is set on seasoning a salad, you might incorporate the defrosted herb into a salad dressing.

It's a good idea to freeze herbs in small quantities, each approximately the amount required for a single recipe. Wash the herbs well in copious cold water before freezing, but don't blanch them in boiling water because that will wash away too much of the essential oils. For the same purpose of conserving these volatile oils, don't chop herbs before freezing them. Make the herb packages airtight, wrapping them well in foil or plastic wrap or bags. Squeeze the air out of freezer bags.

Garlic can also be frozen. Break up the head, but do not remove the papery skins from the individual cloves. The garlic cloves can then be removed from the freezer one at a time, as needed. The clove will become mushy when it defrosts, but it will be a flavorful mush, and for many dishes this texture is perfectly acceptable.

When you are making a dish that you intend to freeze and reheat later, reduce the amount of herbs by about one-third, as the flavor of most herbs intensifies with this process. If you find the dish a bit bland when you serve it, add a sprinkling of fresh herbs for taste or a garnish of sprigs for a full aroma.

CHOPPING HERBS

Whether they come from the supermarket or the garden, herbs should be rinsed well before using. Discard the tough stems, along with any wilted or discolored leaves. Blot them with a towel and let them air dry for a few minutes; herbs need to be thoroughly dry before they are chopped. A salad spinner can be used to speed the drying process.

Not every herb takes well to chopping; some are too small or too fine to attack with a knife. When using only a few **basil** or **coriander** leaves, it is best to tear them into pieces with your fingers. Most varieties of **thyme**, **marjoram**, or **oregano** have tiny leaves that can be simply stripped from their stems into the pot. **Chives** are never chopped, but are snipped with scissors or sliced across into pretty little rings. And why would anyone want to destroy the lovely lacy edges of tender young leaves of salad burnet? Use them whole! Serving a dish with a sprig of fresh **rosemary** is often more effective than chopping the leaves up and adding them to the food; the scent of the leaves will influence the flavor quite sufficiently and tough rosemary leaves are not pleasant to chew.

In contrast, **parsley**—both the curly and flat-leaf varieties—can be chopped as fine as you like, and in some circumstances can even be subjected to the food processor. Note that a food processor and a handheld knife produce different results. The processor tends to make the parsley wetter and a little mushy, especially when you are using the flat-leaf variety. Sometimes this is perfectly acceptable, but it is not recommended, for example, in an authentic Middle Eastern tabbouleh.

A pleasing, regular effect is obtained by making a *chiffonade* of herbs—that is, finely cut ribbons. This is an attractive way of serving quantities of herbs that have larger leaves, such as **basil**, Cuban **oregano**, **perilla**, and some **mints**. The resulting leafy heap serves as an aromatic bed for meats and other foods, or as a garnish, and also as a tasty ingredient in stuffings or in stir-fried dishes. A very fine chiffonade can be used in soups or salad dressings.

To make a chiffonade, arrange the leaves in a neat stack, stems together, then roll the leaves from edge to edge into a tube with the stems sticking out at one end. Slice across the roll with a large, sharp knife to make fine shreds. Discard the stems. Make the thinnest possible slices if your chiffonade is to be used as a flavoring, slightly thicker for a garnish, and cut slices about one-eighth inch thick when the chiffonade will serve as a bed. This same technique works well with other large leaves, such as lettuce and sorrel. Sometimes a chiffonade is simmered in butter, olive oil, or broth for a minute or two until just soft.

TOASTING SPICES

Many whole spices and seeds will generously repay a light toasting just before they are used. Simply place them in a single layer in a dry skillet and heat them over medium-high heat, shaking the pan or moving the spices around with a wooden spoon until you get the first whiff of their wonderful fragrances. Remove them from the pan quickly so they do not burn. When they have cooled, grind them in a mortar or a spice mill. (A small electric coffee mill that you do *not* use for coffee works fairly well for this purpose.) It is best to toast different spices separately, because they may require different toasting times; however, they can all be ground together.

Saffron threads should also be toasted before they are pulverized and added to the dry ingredients of a dish, or steeped in a warm liquid. Because these delicate stigmas and styles can burn so easily, it is safest just to put them on an ovenproof saucer set over a small pan of boiling water until they are crisp. They can then be crushed easily in a mortar or with the back of a spoon; if your recipe is for a sweet, you can add one-fourth teaspoon of sugar, whose granules help cut up the threads as you crush them. Add a little salt for the same purpose if you are making a savory dish.

MEASURING

Measure herbs *after* you have chopped them. Do not pack the chopped leaves in the measuring spoons, just press them down lightly.

Remember never to dip a damp measuring spoon into a jar of dried herbs or ground spice. The moist seasoning will clump together, and is at risk for mold. For the same reason, shaking an herb or spice out of the bottle over a steaming pot is to be avoided—even if the manufacturer has put a shaker top on the bottle. Shaking results in wildly imprecise measurements, and sometimes the seasoning comes out much faster than anticipated. It's better to remove the shaker top and pour the desired amount of seasoning into the palm of your hand first. If dealing with a dried herb, crush the leaves leaves slightly between your hands or with your fingertips as you add them to the pot.

Use the flat back of a knife blade to level off the spices in a measuring spoon for exact measurement, or "level spoonful." A "heaping spoonful" is as much as the utensil will hold; a "rounded spoonful" is somewhere between heaping and level quantities. For a "scant spoonful," the spoon is not quite full. A "pinch" of spice is officially defined as one-eighth teaspoonful, but the term does not imply that kind of accuracy: what you grab between your thumb and index finger is just about right.

"FRESHENING" A RECIPE

Replacing the dried herbs called for in a recipe with fresh leaves of the same herbs will usually improve the flavor of your dish, but you will need to use a larger quantity of fresh herbs. The usual ratio of fresh herbs to dried ones is three to one; which means, for example, that one tablespoon of chopped fresh herb is approximately the same strength as one teaspoon of the same herb in dried form. But there are exceptions to this rule of thumb: for example, a mere half teaspoon of dried **basil** is equivalent to one tablespoon of fresh, while one and one-half teaspoons of dried **dill** are needed to match a tablespoon of the fresh leaves in flavoring power. Thus, to freshen a recipe calling for a teaspoon of dried basil, use two tablespoons of the fresh herb; and if a recipe is content with a teaspoon of dried dill, then only two teaspoons of fresh dill will substitute for it.

As much as possible, alter the recipe to add fresh herbs near the end of the cooking process, which helps them conserve their essential oils. Dried herbs, on the other hand, need to go into a dish early enough to soak up its liquids and rehydrate.

Please note that freshening your recipes is not right for every herb. In the case of **bay leaves**, there is really nothing to be gained in replacing a dried bay leaf with a fresh one, since these leaves retain their flavor very well when dried and, in fact, leaves that have dried just to the point of being brittle are actually preferred. And fresh **mint** should not be substituted for dried mint, as the two have rather different flavors. Similarly, fresh and dried **ginger** are like two separate seasonings, and neither can be used in place of the other.

Using whole spices instead of commercially ground powders also has an amazing freshening effect. Measure out the same amounts of freshly ground spices as called for in the recipe—and just let them sing! Black **pepper** that you crank out of your own pepper grinder, for example, has no more bite than stale preground powder, but it has much more of that heavenly aroma which flew away from the store-bought powder long ago. Grinding your whole spices with a small mortar and pestle generally yields finer results than a mill, but much depends on the mill and on your patience. Whole **nutmegs** should be rasped on a little grater, and **cinnamon** sticks, too tough to grind at home, are best simmered whole in a liquid.

ADJUSTING SEASONINGS

With few exceptions, you can play with the seasonings in a recipe to your tongue's content. If you don't like **anise**, substitute **mint**. If you just adore **paprika**, shovel it in. If you think **five-spice powder** would be good in your dish, go for it! These substances will not alter the cooking chemistry or make your recipe fail.

Exceptional substances are primarily sugar, **salt**, and **chocolate**. Sugar does far more than sweeten; it can also thicken, harden, and allow browning. There is also a limit to the adjustments you can make in the quantity of salt used. Adding herbs, spices, and other flavorings can compensate for salt in many recipes, but often a little salt is essential; salt-free bread is a tricky special case in baking. Chocolate almost requires incantations along with correct temperature and humidity ranges; and it matters very much whether you use cocoa or baking chocolate. Consult the "Individual Seasonings" chapter for details.

It is a fine idea to taste your dish as you create it, but it is not always possible to do so. In particular, you should not taste raw eggs because of the risk of salmonella poisoning. Instead, rely on your nose; the aroma of a dish will tell you a great deal about its flavor. In any case, many flavors, and certainly combinations of flavors, change when they are cooked, so tasting them in the raw state does not help you much in adjusting the seasoning. Note that a sip of the liquid in a soup or a stew—don't burn your tongue!—will tell you more about the spicing than a bite of the vegetables or meat.

If you are doubling a recipe, do not double the seasonings, but use the same amount called for in a single recipe; then taste as soon as feasible and increase the spicing in small increments. When halving a recipe, halve the seasonings; then taste and adjust them gradually. It is infinitely easier to add than to subtract a flavor. If your dish is too salty, put in a little sugar, half a teaspoon at a time.

For a final check on the seasonings, always test your dish again, as close to the end of the preparation as possible. Use your nose to gather information about the flavor if for some reason it is not possible to take a bite. Many times a dish that threatens to be bland and unexciting can be rescued with a garnish of chopped **parsley** or **cilantro**; a sprig of **rosemary**, **mint**, or **dill**; a **scented geranium** leaf; a sprinkling of **sumac** or coarse **salt**; a dusting of **cinnamon**, cocoa, or confectioner's sugar; a **chocolate** or **lemon** sauce; a topping of toasted **sesame** seeds; or a seasoned salt to serve as a dip. The possibilities are virtually limitless; your own seasoning savvy and personal preferences will guide you in selecting one for the occasion.

LEAVENINGS

No, the various kinds of leavening agents, used to raise and lighten doughs and batters, are not actually seasonings. Yet your choice of leavening will have a definite effect on the flavor of the finished product. Some foods are, by their nature or by tradition, associated with certain leavening agents: breads with yeast, for example, or cakes with baking powder, gingerbread with a combination of molasses and soda, crackers with ammonia, and meringues with air. But these customs are not mandatory and even when you follow tradition there are still many choices to make.

Yeasts, and baking powders and other chemical combinations, as well as plain old air and liquids that turn to steam, are the most common examples of substances which will make a food light or even ethereal. Whichever leavening you are dealing with, make it a principle to use the least amount possible to do the job. A tiny bit of yeast will raise a whole loaf of bread if it is given time and the right environment; the taste of baking powder can easily overpower the delicate flavor of a biscuit; and even air can be incorporated to excess. Consider the bubble: Like a balloon, it consists of both an expanding gas of some sort *and* a stretchy substance to hold it in. Most often in baking, this substance is the elastic gluten in wheat flour (although egg whites also serve at times). This double nature of a bubble is why overleavening can actually cause a cake or other food to fall, because too much leavening will produce more expansive gases than the gluten can contain, and all your bubbles burst.

Yeast

Yeasts are everywhere. These microscopic single-celled organisms are found in the air on dust particles and in the soils of every land, *not* excluding Antarctica. They cling to plants and animals and are especially abundant wherever there is sugar. Yeasts are actually minuscule fungi of many different types; not all of them are beneficial to us, but some are the boon of brewers and bakers.

Yeasts of the species *Saccharomyces cerevisiae* perform the magic of fermenting sugar into alcohol and carbon dioxide. Brewers want more of the alcohol and bakers more of the gas (to stretch the elastic gluten in the dough), and various strains within this species are better suited to one or the other of these tasks. Each manufacturer of baker's yeasts works with its own particular strain, and it is worthwhile trying several different brands to find the yeast that is best for your climate, baking habits, and taste. Lately, many bakers have been playing with yeasts, scraping them off wine grapes or taking them from the air with a lump of warm dough. This experimentation can be a lot of fun, but remember that there are no guarantees as to what yeast you will get, how strong it will be, or how it will taste.

Yeast for baking may be purchased in compressed cakes of fresh yeast, or in packets or jars of dried yeast granules; or you may use a lump of "starter" dough set aside during a previous baking. This starter, like the compressed fresh yeast, should be wrapped and stored in the refrigerator, where it will stay alive for about three weeks. Dried yeast must be kept dry and not too hot; under the right conditions, it can survive for several years. Some bakers feel strongly that fresh yeast is superior to dry, but the most important factors seem to be how much is used and how it is handled.

Handling yeast requires moderation in all things: not too much heat, not too much oxygen, not too much moisture, not too much sugar. Yeast will thrive in a warm environment around 80°F; it will die in heat around 120° to 140°F, and at 45° activity practically ceases—although it remains alive at cold, even freezing, temperatures, in a dormant condition. Although yeast feeds on sugar, mixing a lot of pure sugar into a dough interferes with the development of gluten; the usual way to add sweetness to yeast breads is with an icing or a sauce, or by enriching the dough with dried or candied fruits.

Although human beings have been making use of yeasts for millennia, the nature of these microorganisms was not understood until quite recently. Consequently, the available yeasts were unreliable at best, and required "proofing" to make sure they were alive and would start kicking when they encountered warmth and moisture. Today, unless your yeast is outdated or has been stored improperly, it almost certainly is ready to activate, and *proofing* has come to mean simply dissolving the yeasts in warm water.

Dissolving the yeast in a liquid, usually the first step in creating a dough, is unnecessary for the new "instant yeasts." These granules can be added directly to the dry ingredients and proofing can be skipped altogether.

Bakers agree that yeast works best in smaller amounts over longer periods of time. If you try to hurry the rising by adding extra yeast, or push it by warming the dough too much, the resulting baked goods are dry, almost stale, not as tender, and have an inferior taste. Yeasts vary too much to give hard and fast rules about proportions—and the conditions in your kitchen provide other variables—so the amounts specified by manufacturers, or by your recipe, may well be excessive. Try experimenting with reduced quantities of yeast, and allowing your doughs more time to rise. Your patience will be rewarded with a fine, moist, flavorful yeast bread or cake.

Baking Powders

Baking powders depend on the simple chemical principle that when an acid combines with a compound containing some form of carbonate, the gas carbon dioxide is released in the process. It is this gas that causes breads or cakes to rise. Many recipes take advantage of this principle by calling for a combination of acid and a carbonate salt in the dough or batter; usually baking soda (sodium bicarbonate) is chosen to combine with an acid such as buttermilk, sour milk, molasses, or cream of tartar. Only a little baking soda is needed to do the job; a good rule of thumb is that one-fourth teaspoon of baking soda, plus an acid, will perfectly leaven a whole cup of flour. It is also important to measure these substances carefully so that the chemical reaction is balanced and does not leave a part of the soda or the acid in the food, which badly affects the taste.

About a century and a half ago, commercial baking powders were invented to provide a balanced mix of chemicals, ready to react in the presence of heat and moisture. Baking powders are classified according to which acid is used in their reaction; this is usually calcium acid phosphate or sodium aluminum sulfate, and many brands contain a combination of the two. Tartrate baking powders also exist, but are rarer. Almost all types use sodium bicarbonate as the base, but there are health-food brands that substitute potassium bicarbonate for the benefit of those who wish to avoid sodium. A quantity of inert starch, usually cornstarch, is also included in all baking powders. The "double action" promised on the label refers to the separate chemical reactions from the different acids in the combination types of baking powder. However, the other types of baking powders can also claim to be double-acting by virtue of the initial release of gas when moisture is added, followed by the expansion of the bubbles in the heat of the oven.

During the relatively brief history of commercial baking powder, the many manufacturers of these products reacted to each other in fierce and fizzy competition, and ultimately involved the United States Federal Trade Commission in court proceedings for years.

Doctors, chemists, dietitians, home economists, and others all testi-
fied for or against claims and counterclaims that baking powders of
different types were harmful to health. In the end, no baking powder
was officially declared unwholesome, and they are all on the market
today. But although the controversy about them has died down,
there are still many differences of opinion as to which kind is best
for health or flavor. Many older cookbooks even recommended
adjustments in the recipe according to the type of baking powder
chosen.

You can mix up your own baking powder by combining one
tablespoon cream of tartar and one teaspoon baking soda with two
teaspoons cornstarch. Keep this mixture tightly sealed in a dry
place, and use it within a couple months. The chemical reactions in
homemade baking powder begin as soon as moisture is added; so
for best results, get your dish into the hot oven as soon as possible
after the batter has been mixed.

Baking powders have been developed to provide a quick and
easy leavening for cakes, muffins, some cookies, and many other
oven-baked goods, but they are not recommended in waffles, pan-
cakes, flapjacks, or other stove-top goods. Instead, create your own
chemistry with soda and sour milk or other base/acid combinations.
As a general rule, one cup of sour milk or buttermilk of average
acidity will neutralize one-half teaspoon of baking soda. Incidentally, this is an ideal amount of leavening for two cups of flour.

Ammonia

Baker's ammonia (ammonium carbonate) is another alkaline
substance, like baking soda, that has been used in creating chemical
leavenings. Older recipes often call for it as "hartshorn" or "hartshorn salt," in reference to a traditional source of this substance.
Ammonia powder is generally dissolved in a liquid before being
incorporated into the recipe, rather than being added with the dry
ingredients as baking powder is. For best results, use a low oven
temperature and a long baking time for ammonia-raised goods in
order to bake off all the gas.

Ammonia is popular in Greek, Scandinavian, and German baking, and is especially good for flat products such as crackers, biscotti, lebkuchen, and other baked goods that you want to be very crisp.

Don't worry about its penetrating pungent smell; at the end of the full baking time, this will be entirely gone.

Air

What better way to aerate a dish than with air itself? This may be the only flavorless substance discussed in this book, but the absence of taste is itself an important aspect of flavor.

Angel food cakes, meringues, and soufflés are raised simply by air trapped in multiple pockets of egg white and expanding in the heat of the oven. A practiced cook with a balloon whisk can make the whites stand at attention in stiff peaks faster than an electric beater can do it, but either one will get the job done. Of course, any trace of egg yolk or fat will defeat the effort. The cloud of stiffly beaten whites can be stabilized with a little acid, either a pinch of cream of tartar or a small squirt of lemon juice added after the egg whites are frothy, or by reaction with the metal when they are beaten in a shiny copper bowl. (Of course, such a copper bowl cannot be kept shiny with poisonous metal polishes; lemon juice and salt do the job well and safely. Slice a lemon in half, lay a teaspoon of salt over the cut side, then rub it vigorously all over the bowl.) Sugar also helps to stabilize the whites.

Egg whites can be overbeaten, and most recipes tell you to beat them until "stiff but not dry." If you beat beyond this point, you must watch carefully to be sure the foam still looks glossy and moist. Overbeaten whites are stretched to the limit, and will collapse in the oven.

Beaten biscuits are also raised with air, a successful attempt to eliminate the distracting taste of baking powder from these light, tender, hot and heavenly, soft-wheat Southern treats. A simple mixture of flour, water, fat, and salt is laboriously beaten until blisters form in the dough. It is the expanding air in these blisters that leavens the biscuits as they bake.

Steam

Steam expanding in a hot oven is always a help to any form of leavening, but in a few cases steam does the work all by itself. This happens, for example, with Yorkshire pudding, popovers, and *pâte à choux* for cream puffs and eclairs.

CITRUS

The *Citrus* genus is wildly prolific, adapting to different climates and conditions to produce fruits varying in color, size, shape, and flavor. These fruits include the citron, grapefruit, **lemon**, **lime**, **orange**, pomelo (also called shaddock), tangerine, the Japanese *yuzu*, and all their numerous varieties and hybrids. Several of these fruits are quite new, botanically speaking, having been developed only in the past century or two, and horticulturists are at work at this very moment creating yet more varieties. A few dictionaries and encyclopedias expand the term *citrus* to include fruits in other genera, such as the little oblong kumquats, which belong to the genus *Fortunella*. But kumquats, with their thin, sweet, edible rinds and sour pulp, do not comply very well with the following generalizations, and in this book citrus fruits are strictly defined as those belonging to the *Citrus* genus.

Although their appearance and flavors are stunningly different, the citrus fruits do have some common traits: fragrant blossoms; shiny, evergreen leaves; and fruits having an aromatic, oily skin with a pithy underside, and a juicy, acidic interior. (All of these plant parts, from one citrus species or another, are used as flavorings.) Thus, certain culinary techniques are suitable for all of these fruits. (See also **lemon**, **lime**, and **orange** in the "Individual Seasonings" chapter.)

Selection

Although you can get fresh citrus at any time, the winter months offer the best selection. Take advantage of it. Citrus fruits should neither be shriveled nor have soft spots, but should be firm and feel weighty in your hand. A fine-textured smooth skin, with small pores close together, indicates a juicy fruit, while knobbly, large-pored skin means it has a thicker rind and less juice; choose the one that suits your purposes.

Storage

Store citrus fruits in the vegetable compartment of your refrigerator. Good fresh fruits should last for several weeks.

Peel

When you intend to use the peel, or rind, of citrus, always wash the fruit well and scrub off any inky stamps. It is advisable to choose organic fruit for this purpose to be sure that the rind contains no pesticides or dyes.

Citrus rind is made up of the zest—the outer, colored part—and the inner white pith. The zest contains the wonderfully aromatic essential oils. The pith, on the other hand, can be rather unpleasantly bitter, so many recipes call for the zest only.

There are several ways to remove the zest from the fruit. It may be lightly scraped off with a fine grater; do not continue grating in a spot where the white pith shows through. In a delicate smooth custard, sweet soufflé, or mousse, however, you may prefer not to have chewy little bits of zest. For these dishes, you can pick up a surprising amount of the essential oils by rubbing a sugar cube over the surface of the citrus fruit. A cube rubbed vigorously over the skin of a fruit until all six of its sides are slightly colored will contain nearly the same amount of essential oils as one-fourth teaspoon of grated zest. The sugar cube is then simply added with the rest of the sugar in the recipe.

There are numerous gadgets for removing zest, some of them yielding fine, long threads. You can also use an ordinary vegetable peeler to slice off thin strips of zest, taking care to avoid digging too deeply into the pith. Julienne the wider strips that were cut with the vegetable peeler; if your recipe requires small bits of zest, then cut across the julienned strips at right angles. If you want the zest milder in flavor and softer in texture, then blanch it in boiling water, drain, cover with cold water, and boil it up again for about three minutes.

Once the zest is off, the fruit should be wrapped in plastic or placed in a tightly sealed container to keep it from drying out, then stored in the refrigerator. Later, it can be eaten or juiced—but don't wait too long because it will not keep as long as intact fruit would.

The rinds of citrons, grapefruit, and the thicker-skinned varieties of oranges and lemons are especially suitable for candying. If you are using more than one kind of fruit, cook the peels separately to maintain their distinctive flavors.

To prepare the candied peel, slice off the stem end of the fruit and score the peel longitudinally from pole to pole with a sharp knife, cutting just to the pulp and slicing the peel into four or more pieces. Pull off the peel, leaving as much of the membrane on the pulp as is possible. The more bitter peels—citron, pomelo, grapefruit, and bitter orange—are usually soaked overnight in salty water as a first step. Use about one teaspoon of pickling **salt** or kosher salt to one cup of water for soaking. Finally, rinse the peel well.

Again slicing longitudinally, cut the peel into smaller pieces of regular size and shape, about half an inch wide at the middle and tapering to a point at each end. Put the pieces into a saucepan, cover with fresh, cold water and bring to a rolling boil. Then drain the peel, cover with cold water again, and bring to a boil a second time. Repeat this procedure once or twice more. At this point, the peel should be tender; if it is not, then let it boil a while longer. The peel must be soft before it is candied; otherwise the sugar will make it tough. Finally, drain it thoroughly in a colander.

Measure out a quantity of sugar approximately equal to the amount of peel (this may be by volume or by weight) and return both to your saucepan; just cover the rind and sugar with cold water and bring the mixture to a boil over medium heat. Let it bubble gently—don't cook it too fast—until the fruit has absorbed much of the sugar and has become somewhat translucent. Remove the peel from the syrup, roll it in granulated sugar, and lay it on a rack to dry. If the humidity is extremely high, dry the peel on the rack in a very low oven.

The colorful, jewel-like strips of citrus peel that result make a beautiful and delicious candy in themselves, or they can be cut up and added to fruitcake, stollen, panforte, marmalade and preserves, rice pudding, and many other confections. Store candied peel in the freezer. It will keep indefinitely.

When you halve a citrus fruit at the equator and carefully scoop out the pulp with a spoon, the remaining hollow shell makes an attractive container for many foods. If you prefer, cut the fruit in zig-zags to give the rind a pinked edge; you may also have to cut the ends flat to keep the hemisphere upright. Orange-rind bowls can hold chutneys and relishes on a platter of meats; hollowed-out lemon halves are perfect for a dipping sauce for shrimp; and lime shells are ideal to hold individual servings of spicy salsa made with chiles. Consider lemon sherbet served in frozen lemon shells; citrus and curly endive salad in a hollowed-out grapefruit; or hot mashed sweet potatoes, flavored with a little orange juice, in an orange cup. Candying any of these shells will make them look prettier and last longer. Simply follow the same procedure as for candied strips.

Dried (not candied) tangerine peel is a popular flavoring in Chinese cuisine. Strips of the hard peel are soaked in water until they become pliable, then the thin layer of bitter pith is scraped off the underside, and the peel is ready to simmer in a savory dish.

Persian cooks are also fond of dried citrus peel as a seasoning; sour oranges are traditional, but sweet oranges or tangerines are also used. They remove the pith and cut the peel into fine slivers before drying it. Just before using, a handful of shredded peel is boiled up two or three times, always beginning with fresh cold water, to remove the bitterness.

Juice

Citrus fruits yield less juice when they are cold. To get the maximum amount of juice from these fruits, heat them in a microwave oven before cutting them. Zap a lemon for fifteen seconds on the high setting, and larger fruits just a little longer. Then juice as usual, with whatever gadget you prefer; note that the type of juicer that squeezes an entire citrus hemisphere in a vice also expresses some of the tangy oils in the skin as well as some of the bitterness of the pith and seeds. Of course, a citrus fruit can be juiced simply by stabbing the tines of a fork into the cut surface of the pulp and twisting; hold the fruit over a sieve to catch the seeds. If you are serving a half lemon or lime on a plate, for the diner to squeeze, the seeds can be caught by dressing the fruit in a bit of cheesecloth or net. Place the fruit cut side down in the middle of an ample circle or square of cloth, then tie the edges up tightly at the point of the fruit with a bit of string or ribbon. For an attractive presentation and an intriguing aroma, tie into the knot a sprig of fresh **mint, parsley, dill,** or other herb.

Bottled citrus juices are convenient, but their flavor is simply different from that of the fresh juices.

Pulp

The pulp of citrus fruits is naturally divided into individual segments, each enclosed in a thin membrane. This allows an attractive presentation for some dishes, and makes for neatness when eating with fingers or toothpicks. In many cases, however, it is better to remove the membrane around the pulp. Slice off the peel with a sharp knife, taking care to remove all traces of the white pith under the rind. This is messy work, best done over a large bowl or a sink. You can then slice the fruit latitudinally into beautiful rounds that display lines of membrane radiating starlike from the center. These citrus rounds are particularly effective in composed salads.

When no membrane at all is desired, as for ambrosia and other fruit desserts, the pulp can be lifted out, section by section, by first removing all white membrane on the outside of the fruit and then slicing down to the center along both sides of each separating membrane to liberate the segments. Hold the fruit over a bowl to catch the juices as you perform this operation.

Leaves

All citrus leaves are edible and can be used as seasoning, although the amount of essential oil they contain varies from species to species. Kaffir lime leaves are the most potent, but every citrus leaf is aromatic. Shiny green citrus leaves of all kinds are excellent in fruit punches or as a garnish for fruit desserts. If you're out of lemons for your tea, but have a supply of citrus leaves in the freezer or on the tree, then the leaves will substitute nicely for the fruit; use one whole leaf for each cup or glass.

Citric Acid/Lemon Salt/Sour Salt

The sour taste of citrus fruits is due to the presence of unusually large amounts of citric acid, an organic acid essential for metabolism in both plants and animals. It forms odorless white crystals with a very sour flavor, which dissolve readily in water and other liquids, including oils. Citric acid crystals can be bought at the drugstore or at Middle Eastern, Indian, and other ethnic groceries.

Citric acid is used to prevent the discoloration of foods that results from oxidization, to add a sour punch to a dish, or to increase the acidity of foods for safe canning and preserving. Notice how often citric acid appears among the ingredients listed on the labels of prepared foods and beverages.

One teaspoon of the crystals is approximately equivalent to the juice of one medium lemon in acidity or sourness—or, more precisely, about eight teaspoons of juice—but it does not add a lemon flavor. You can use citric acid in place of lemon juice in jams and watermelon rind or other preserves. It is important to use it in putting up some of the newer varieties of sweet tomatoes, which otherwise may not have sufficient acidity to discourage botulism.

In India, citric acid is sometimes used to curdle milk in preparation for making *panir*, the traditional Indian cheese. Mexicans sprinkle a mixture of ground **chiles**, **salt**, and citric acid over fingers of jicama, pineapple or cucumber wedges, or round slices of peeled oranges. These make up an important part of the popular Mexican snacks called *antojitos* ("little whims") that go so well with alcoholic drinks and fruit juices.

TEAS AND TISANES

A tea, or to use the French term, a *tisane,* is an infusion of flavorful plant material in hot water, below the boiling point. While the best-known tea herb is *Camellia sinensis*, yielding green teas, black teas, and oolong teas, many other plants, including herbs and spices, can also be brewed into delicious tisanes or herbal teas. Bringing out the best qualities of any kind of tea, however, requires some attention. Moreover, the technique differs according to the material you choose to brew.

The following method is recommended for leafy herbs, such as **mint** or **wintergreen**, and for flowers or flower parts, such as **monarda** or **saffron**. First of all, begin with fresh cold water. Running tap water, preferably filtered, is usually best, but if your tap water does not taste good, choose bottled water instead. Bring an ample quantity of water to a full boil in a kettle. Do not overboil the water, for it will lose its aeration and the tea will taste flat.

If possible, choose a ceramic or glass teapot; metal pots often impart a taste. Warm the pot with a little hot water from the kettle, then pour it out. Now add approximately one teaspoon of dried herb for each cup you want to make. The exact measurement depends upon the type of herb selected, its freshness and strength, and your own preferences.

The rule is to bring the teapot to the boiling kettle, rather than the kettle to the pot, to ensure that the water is properly hot. Pour the water over the leaves, cover the pot, and let the tea steep for three to five minutes. Again, the steeping time depends on the herb you are using. If you want to make the beverage stronger, use more herb rather than more time because the best flavor generally steeps out first.

When the tea has steeped, pour it through a strainer into pretty cups, and serve it very hot. Stir in sugar if desired; for some herbs, such as **balm** leaves, honey is a more harmonious sweetener.

If you find your herbal tea is too strong, you can dilute it with a quantity of hot water. But if it is too weak, you cannot rescue it by adding more leaves at this stage. Instead, try adding a little extra flavor with lemon slices or milk.

Elegant decoctions may also be brewed from various spices, but the method differs somewhat from that for leafy herbs. Spices need a little more time to give up their essences in an infusion. When dealing with seeds, such as **dill**, or fruits, such as **rose** hips, crush the spices slightly, then cover with fresh, cold water in a pot; bring to a boil, then turn down heat and simmer for about eight minutes or longer, until the flavor is sufficiently strong for your taste. To make beverages from roots, such as **licorice**, or barks, such as **cinnamon**, more simmering time is needed, perhaps as much as ten or fifteen minutes.

Individual Seasonings

Individual seasonings appear in this chapter in alphabetical order. Alternate names are listed in the Index. Seasonings and blends appearing in **bold** type have their own entries in the book. See the following chapter for flavor combinations of all kinds.

A **ajowan/ajwain/bishop's weed/omam/carom seed/netch azmud** (*Trachyspermum ammi*): The ribbed, dry fruits (generally called seeds) of this herb are less than one-eighth of an inch long, but if you look closely you can see their longitudinal stripes of red-brown and tan. Although small, these fruits are definitely assertive. They have a pungent, somewhat rough aroma, and when chewed raw they sting the top of the tongue; however, their taste and smell mellow delightfully in cooking.

Ajowan is an important seasoning in both Indian and Ethiopian cuisines, and its unique flavor makes it worth looking for when you are cooking dishes from these countries. It is relatively easy to find in Indian shops.

Ajowan has an affinity for beans of all kinds, and is believed to help with their digestion. It also marries well with seafood, and many Indian cooks consider it absolutely essential for fish curries. Try a little in a spicy marinade for fish. Ajowan is a flavoring for *nan* and other Indian breads. It is also used in Indian pickles, in a vegetable filling for *pakoras*, and as an ingredient of **chaat masala**.

Ethiopian cooks put ajowan into many of their meat stews, and use it in making a wheat bread. It is also included in the hot Ethiopian **berbere** seasoning, which is a basic ingredient for many other dishes in that cuisine, such as the famous chicken dish *doro wot*.

Ajowan should be toasted lightly (see Toasting Spices in the "Culinary Practice" chapter) and ground before use, unless your recipe has you sizzle the spice in hot fat for a few seconds.

Ajowan seeds and the herb **thyme** both contain a great deal of the essential oil known as oil of thyme, so you can substitute equal amounts of dried thyme for ajowan if necessary. (Of course the thyme leaves should not be toasted or fried.) However, a better approximation of the penetrating, lingering flavor of ajowan is achieved by a mixture of thyme, **cumin** seed, and **celery** seed; in place of one teaspoon ajowan, mix one-half teaspoon dried thyme with one-fourth teaspoon toasted whole cumin and one-fourth teaspoon lightly toasted celery seed.

allspice/pimento (*Pimenta dioica*): Allspice, despite its name, is a single spice from a single source. The allspice berry fruits on a tall evergreen tree that seldom grows well outside the Caribbean and Central America. Jamaican allspice is by far the finest grade, and Jamaica has long dominated the world market in this spice. For these reasons, allspice is sometimes called Jamaica pepper. On that island, and in the spice trade, the spice is known as pimento, a name derived from the Spanish word *pimienta*, meaning pepper. Our name "allspice" refers to the fact that the flavor resembles a combination of **cloves**, **cinnamon**, and **nutmeg**.

The whole dried allspice berry looks like a smooth, plump, reddish-brown peppercorn. If you are uncertain whether you have a large peppercorn or an allspice berry, your nose will tell you which you are dealing with; if it does not, do you have a cold, or is your spice too old? If you cut an allspice berry in half and look inside, you find two small seeds. (A peppercorn, in contrast, contains one large central seed.) When you shake a good-quality, properly dried allspice berry near your ear, you can hear those two seeds rattling inside.

Allspice is most often used in sweets, such as pumpkin pies and spice cakes, but it is also an excellent spice for meats of all kinds. It is much used by the American food industry in producing sausages, cured meats, and cold cuts. Ground allspice is wonderful in Swedish meatballs, and the whole berries are attractive and flavorful when floated in a cream sauce for this dish. Allspice is an essential ingredient in the popular Caribbean **jerk seasonings** for pork, beef, or chicken.

Whole allspice berries are standard ingredients in **pickling spice**, and they are especially important in cured fish such as rollmops and gravlax. Drop in two or three whole berries per cup of liquid when poaching seafood of any kind.

Allspice is frequently an ingredient in the spice mixture used to make spiced teas and mulled wine or cider (see **mulling spices** in the "Flavor Combinations" chapter). It is frequently used in Indian curries, and it is the most important spice for seasoning rice dishes in Turkey. Allspice is one of the reasons why we love ketchup, and should be considered as a seasoning whenever you are making a tomato-based sauce.

The whole berries can be freshly ground in an ordinary household pepper grinder, provided there is enough room for the whole berries to fall into the grinding mechanism. "Pepper" mixtures, intended for meaty dishes, of whole black and white **peppercorns**, allspice berries, and sometimes green and **pink peppercorns** as well, can be set on the table in a transparent plastic pepper mill. The colorful mixture looks attractive, and when ground it releases a heady aroma of warm spices.

The volatile oils in allspice are especially flighty, so you need to take extra care to store this spice properly (see Storage in the "Culinary Practice" chapter). Buy it in small amounts, and replenish your supply often.

Substitute for ground allspice a combination of ground cloves, cinnamon, and nutmeg, in equal quantities. Conversely, allspice can be used in place of any one of those three spices, measure for measure.

almond (*Prunus dulcis*): Almonds serve many important culinary functions. Whole roasted almonds, salted or spiced, are popular snacks; they may be chopped, sliced, or slivered for use as toppings and garnishes; the ground nuts are excellent thickeners and serve as the basic stuff of almond paste and marzipan; almond oil is used in fine baking for greasing cake pans and molds. But it is almonds' role as a sophisticated flavoring that is of primary interest here.

There are two basic varieties of almond: sweet (var. *dulcis*) and bitter (var. *amara*). The bitter almonds contain amygdalin which, when unheated and mixed with water, can yield poisonous prussic (hydrocyanic) acid, so they should not be eaten raw in any significant amount. This substance is not present in the sweet almond variety. However, the taste of sweet almonds tends to be rather bland, and cooks used to include just a few bitter almonds with the sweet ones to make a more interesting flavor. This culinary technique is not feasible in the United States today, however, as raw bitter almonds are not available on the market. Fortunately, almond extract provides the intense flavor of the bitter type without the toxicity, and a small amount of extract should always be added to the sweet almonds in a recipe.

Almond flavor marries well with fruits; try adding a teaspoonful of almond extract to peach and cherry pies and cobblers. The other way around, an apricot glaze is commonly applied to almond tarts. Just a hint of almond flavor will liven up whipped cream; use one-fourth teaspoon extract per cup of unwhipped cream, adding it midway through the beating.

Ground almonds lend an important flavor to macaroons, and they are often incorporated into pastry dough, especially when the pastry is to have a fruit filling. They may be ground with or without their brown skins, depending on how you want your dish to look.

Removing the brown skins from the nuts is time-consuming but easy. Begin by pouring an ample amount of boiling water over the nuts in a bowl and letting them sit for a minute or two. Then drain them in a colander, rinsing briefly with cold water. The skins will then slip off the nuts with a gentle pinch. (If not, repeat the blanching and soaking.) Dry the peeled nuts on a dishtowel, or toast them lightly in the oven.

Toasting almonds, whether blanched or unblanched, brings out their flavor dramatically. Spread whole nuts in a single layer on a cookie sheet and set them in a 350°F oven for no more than eight minutes; if you are starting with sliced or slivered almonds, limit the time to five minutes. Stir them frequently, and do not let them brown. If you intend to grind the nuts, cool them thoroughly first.

Both the food processor and the blender work well enough for grinding almonds, but don't overdo the grinding or the ground nuts will clump up. If this happens, add about a tablespoon of flour or sugar, depending on which suits your recipe; this will soak up some of the oil and keep the ground nuts from forming into lumps. A meat grinder will express the oil from the nuts and should be used only if you want to make almond butter.

In the Middle Ages, European cooks used vast quantities of almonds, and "almond milk" was a kitchen staple. This delicious liquid is made by simmering blanched, chopped or coarsely ground almonds in water. Naturally, the quality of this milk varies according to the proportion of nuts to water, but you can get excellent results by simmering about half a pound of almonds, blanched and finely chopped, in two cups of water for ten minutes; the resulting liquid will have the consistency, but not the flavor, of milk. Strain out the nuts and save them for another use. Try this time-honored ingredient in place of milk or cream the next time you make a delicate almond cream or custard, or an almond-flavored *crème brûlée*; this "milk" will really give your dish almond flavor! Don't forget to add a teaspoon or so of almond extract to supply the intense bitter almond taste: start with one teaspoon of extract per cup of almond milk, then add a few drops more according to your taste. Almond milk needs to be refrigerated, just like cow's milk.

If you want bits of almond for a crunchy topping, chop them with a knife rather than using a machine. And be sure to toast them first. A good substitute for almonds, when used as a topping, is toasted **sesame** seed.

amchur/amchoor/mango powder (*Mangifera indica*): Made from sun-dried unripe mango, this powder is popular in Indian cuisine as a flavor-brightener for vegetable dishes that otherwise tend to be a little bland. Since the fruit is still unripe, the flesh has not attained its full fragrance nor its intense golden-orange hue. Thus, amchur is a pale beige color and not very aromatic, but it adds a wonderful sour flavor with a tropical-fruit tang. It is used with potato pakoras, or added to vegetable side dishes such as spiced cauliflower, eggplant, or okra, and vegetarian curries. It also flavors *chaat*, the popular Indian snack foods often sold on the street (see **chaat masala**, under Herb and Spice Blends in the "Flavor Combinations" chapter).

The East Indians who settled in the West Indies brought with them a fondness for amchur, and among their contributions to Caribbean cuisine are a number of spicy chutney-like condiments made from green mangoes.

Amchur powder is quite delicate and should be added at the end—or as close as possible to the end—of the cooking time. Sprinkle a little extra over the dish just before serving. Amchur should be stored in the refrigerator in a tightly sealed container.

Although this seasoning is usually sold in powdered form (in fact, the name in Hindi means "mango powder"), you sometimes find dried green mango in slices. The slices should be cooked with the other ingredients of a dish to release their flavor, but are best removed before the dish is served.

You can substitute **lemon** juice for amchur; use twice as much lemon juice as you would mango powder.

angelica/garden angelica (*Angelica archangelica*): Angelica is rare among herbs in that its native habitat lies far north of the sunny temperate zones preferred by most other herbs. This tall plant, long known and honored in Iceland, Scotland, Lapland, Finland, Siberia, and Alaska, is not grown commercially in America. You can raise your own angelica if you have a garden in the cooler part of the continent, but the plant does not grow well in the South. There is an American species, *Angelica atropurpurea*, sometimes called masterwort, which can be used in the same ways as *A. archangelica*.

Angelica is best-known as a decorative element in cakes and other confections. When its thick, hollow stems are candied (see instructions below), they turn an attractive vivid green, useful for creating colorful patterns or for making the "leaves" in flower designs made of glacé fruits. A certain amount of candied angelica is imported from France for professional confectioners, especially around Christmastime. Ideally, the green bits in candied fruit mixtures intended for holiday fruitcakes are angelica, but candied angelica is so expensive and rare these days that this niche is usually filled by candied green-dyed cherries or, in some inexpensive mixtures, inferior fruits or even insipid vegetables, dyed green and candied; their contribution to the flavor of the cake is nil. This is an unfortunate loss, for angelica has a unique fresh flavor: somewhat sweet, somewhat bitter, a little musky, and a little resinous.

All parts of the angelica plant are strongly aromatic and edible. The seeds, and the oil from seeds and roots, contribute to the complex tastes of various cordials and digestive liqueurs. The dried roots can be used as a substitute for **juniper berries** in making gin. The leaves and stalks can be used to give zest and interest to custards and ice cream: steep angelica leaves or stalks in the milk or cream called for in the recipe until the liquid has absorbed their flavor, then discard the angelica, and use the cream to make the custard as usual. Whole angelica leaves also impart a refreshing flavor when added to the poaching liquid for fish or seafood.

Tender young angelica stalks are excellent stewed up with rhubarb and other tart fruits such as plums; their angelic sweetness mellows the combination, and you will want to add less sugar than you usually do when stewing these fruits. Use about three times as much rhubarb as angelica stalks, cutting both into pieces of approximately equal size. Add water and a little sugar, and cook over low heat until tender, tasting and adjusting the sugar as needed.

To make candied angelica, cut fresh stalks into three- or four-inch lengths, and wash them well. Blanch by pouring boiling water over the angelica stalks in a bowl and letting them sit for several minutes. Drain, and trim off any tough outer peel. Meanwhile, heat equal parts water and sugar to make enough syrup to cover the angelica completely. Add the angelica to the syrup and let it simmer for fifteen to twenty minutes. Then lay out the stems in a single layer in a shallow pan and pour the hot syrup back over them. Let them stand overnight in the refrigerator or some other cool place protected from dust.

The next day, the syrup will be thinner, diluted by the juices of the angelica. Pour the syrup back into a saucepan and bring it to a boil; cook it down until it is thick once again, and pour it back over the angelica. Let it stand for a second night, and on the third day repeat the procedure.

Finally, on the fourth day, put both syrup and angelica into the saucepan and boil gently for another fifteen to twenty minutes. Saving the sugar syrup for other uses (see below), lay the angelica on a rack to drain, placing a cookie sheet underneath. When the syrup no longer drips, sprinkle the stalks with fine granulated sugar.

The candied angelica must now be dried *completely* to prevent mold. Put it on a rack in an oven set to very low, with the door ajar, for several hours. When dry, store the angelica between sheets of waxed paper in an airtight container or in sealable plastic bags.

The reserved sugar syrup remaining from the candying process is excellent with fruit salads, pies, and cobblers. Use it also to flavor fruit punches, or add a little to orange marmalade; a few fresh leaves or stalks of angelica can also be included in the marmalade to perk up the flavor.

The Iranian spice **golpar** is sometimes identified as angelica seed, but this is an error. These seeds are not at all the same and should not be substituted for each other.

Since angelica is relatively hard to find, consider substituting a few **juniper berries** in such recipes as fruitcake or rhubarb pie.

anise/anise seed/aniseed (*Pimpinella anisum*): Anise "seeds" are really small dry fruits. Brown, ridged, and highly aromatic, they have a sweet, warm taste. Much of their flavor is due to the presence of an essential oil known as anethole, often described as imparting a **licorice** flavor. This is confusing, because generally the dominant flavor in licorice candies comes from anise oil, extracted from the seeds, rather than the milder sweet taste of licorice-root extract. Anethole, in widely varying amounts, also contributes to the flavors of other herbs and spices, including **avocado leaf**, some **basil**, **chervil**, **fennel**, **hoja santa**, **licorice**, Mexican mint **marigold**, **star anise**, and **tarragon**.

Anise seed is sometimes called sweet cumin and in fact anise seed and **cumin** seed are similar in appearance, although the anise is rounder and the cumin more elongated. However, they taste very different, especially in that cumin lacks the licorice flavor; also, anise is indeed sweeter than cumin; the latter has a sharp, sour component missing in anise seed. In flavor, anise seed is more appropriately compared to the larger, lighter-colored **fennel** seed which shares, to an extent dependent on the variety, the licorice taste of anise seed. Here too, anise seed is by far the sweeter spice, lacking the bitter principle found in fennel seed.

Anise seed is generally sold whole because the ground spice does not hold its flavor long. Best results are obtained by toasting the whole seeds gently and crushing them just before use (see Toasting Spices in the "Culinary Practice" chapter). The flavor of anise seed is penetrating, so use this spice cautiously to avoid overpowering the dish.

German and Scandinavian cake and cookie recipes often call for ground anise seed, the traditional Christmas treats *Pfeffernüsse* and *Springerle* being popular examples. Anise is one of the most popular flavors for Italian biscotti, those hard, twice-baked cookies meant for dunking in wine or coffee. Chocolate-coated biscotti illustrate the remarkable compatibility of anise flavor with **chocolate**. Chocolate bonbons with anise-cream centers are another good example of this delightful pairing.

To flavor cream or milk with anise for making bonbons, custard, or pudding, steep toasted, lightly crushed seeds in the hot liquid for about ten minutes, then strain.

Mexicans like anise seed in sweet breads, and also make a delicious, stomach-soothing tea by steeping the seeds in boiling-hot water. Some types of Mexican sausages and sauces, including the famous *mole poblano*, contain anise seed among a host of other seasonings. In India, a generous pinch of whole seeds is often chewed after a meal to help the digestion.

Anise seed is highly recommended in dishes made with chestnuts or with figs; try adding a little bit of crushed anise seed to fig preserves or fig jam. Mix a teaspoon of crushed, toasted seeds with a quarter cup of sugar and sprinkle it over fresh fruit or on warm bread. Include anise seed in the butter and brown sugar filling for baked apples.

A widespread—but by no means universal—fondness for flavoring clear alcoholic beverages with anise seed gives us the Spanish aperitif *anis*, Greek *ouzo*, Turkish *rakı*, Lebanese *arrak*, Latin American *aguardiente*, and various French *pastis* including Pernod, Ricard, and Berger. These drinks are themselves used as flavorings in other dishes: pastis goes in Oysters Rockefeller, and is added to lobster sauce, fish soups, and the garlic butter served with snails.

If you have anise in your garden, you can make use of the leaves. The anise herb has two types of leaves: ignore the broad leaves at the base of the plant in favor of the thin, feathery leaves near the top. Use these aromatic upper leaves in salads, or steep them in vodka or other alcoholic beverages to add a subtle flavor to drinks. These leaves, finely chopped, are delicious with buttered carrots. (This carrot dish is equally good, if less colorful, when seasoned with melted butter enriched with ground anise seed; but use only a few of the seeds, since they have a stronger flavor than the leaves.) Chopped anise leaves can be mixed with cream cheese and spread on sweet crackers or plain cookies for a beguiling snack.

A similarly flavored but totally unrelated spice is the Asian **star anise**. Star anise also contains substantial amounts of the essential oil anethole, so that ground star anise and ground anise seed can be used as substitutes for each other in approximately equal quantities. Both spices should be lightly toasted.

annatto/achiote/atchuete/bija/roucou (*Bixa orellana*): Brick-red annatto seeds, with a staining juice, grow inside beautiful hot-pink seed pods on a small tropical tree of the New World. The plant is sometimes called the "lipstick tree," indicating one native use of it. The color extracted from the seeds is a lovely golden yellow, which is commonly added to butter, margarine, and yellow cheeses; annatto's mild, almost bland, taste blends particularly well with these fats, and it is an entirely natural food coloring.

Cooks of the Caribbean, the Philippines, and Mexico—especially the Yucatan Peninsula, where the plant is probably native—esteem annatto for its delicate flavor and haunting aroma as much as for its golden color. In the Yucatan, pit-roasted chicken and suckling pig (*pollo pibil, cochinita pibil*) are traditionally seasoned with a red paste of ground annatto seeds mixed with other spices. The national dish of Jamaica, salt cod and ackee, is served with a sauce of annatto, **chiles**, and onions. Among the Filipino foods featuring annatto are the peanut-flavored beef stew *kare-kare* and the popular pork and chicken combination *pipian*.

When buying annatto, select reddish seeds; the brown ones are likely to be old.

The traditional way to use annatto is indirectly, by extracting the color and flavor from the seeds in a little hot lard or vegetable oil;

this fat is then kept for use in various recipes. Cook the annatto seeds gently in the oil or lard, testing its temperature first with a single seed: if the seed sizzles, the oil is ready. For one-fourth cup of fat, use one or two teaspoons of annatto seeds, depending on their age; the fresher they are, the more color and flavor they will surrender. Be careful not to burn them with too high a heat, and put a lid on the pan because the seeds will pop. When they have simmered down and the color of the fat is golden yellow, strain out the seeds and discard them. The colored oil or lard can be kept, covered, in the refrigerator for several weeks, but the fresher it is, the better.

If you wish to avoid the fat, you can also use ground annatto seeds. The seeds are extremely hard, so it is best to boil them for approximately five minutes and then leave them to soak a bit before grinding; otherwise you may find uneven coloration and possibly the unpleasant grit of insufficiently pulverized annatto seed in your dish. After soaking, drain the seeds well, then pound them in a mortar or buzz them in an electric coffee grinder. Be careful as you work with them, because their color is tenacious. For best results, the ground seeds should be mixed with a hot liquid during the preparation of your dish.

Ground annatto seed makes a beautiful yellow rice; just add about a teaspoon of the powder to one cup of water and use that water to make your rice in the usual way.

A prepared liquid seasoning of ground annatto seeds and water, generally imported from the Philippines, is also available. Simply stir or shake this up and add it during cooking, a teaspoon at a time, to the other liquids in the dish until the desired color has been achieved. Be careful not to add this water-based seasoning to hot fat, however, because it will splatter dangerously!

Annatto seed is sometimes used as a substitute for **saffron**, providing saffron's color but not its aroma or flavor. Add the annatto oil or annatto water at the appropriate stage in your recipe until the color is right; the flavor of annatto is so mild that you don't have to worry about using too much.

If you're out of annatto seed, you can make quite a good replacement with a mixture of equal parts of **turmeric** and mild **paprika**. Use this just as you would use ground annatto seed; however, do not fry this substitute mixture in oil or the paprika will be burned and bitter.

asafetida/hing/perunkayam (*Ferula asafoetida* and *F. narthex*): The odor of these large fennel-like plants is fetid indeed, due to the presence of the same sulfur compounds found in **garlic** and onions. Many people—but not all!—find this aroma unpleasant and even repulsive; divided opinion has labeled the spice both "devil's dung" and "food of the gods." But the penetrating smell disappears when asafetida is cooked, leaving only a delicate onionlike flavor.

This spice is obtained by slicing the top of the taproot of the asafetida plant to allow its milky sap to ooze out and gradually harden on contact with the air into pearly "tears" of gum resin. The quality of this product depends on the individual plant, its situation, and the weather; in particular, the ratio of gum to resin varies considerably. A particular lump may or may not dissolve well in water. Furthermore, commercial asafetida is mixed with wheat flour or rice flour, as well as other substances such as gum arabic, to make it easier to handle; this of course affects the quality of the spice you are trying to cook with. Yet despite its variable condition on the market you can make asafetida work its magic for you.

Culinary use of asafetida is confined almost exclusively to Indian cuisine. It is especially popular in Kashmir, where it is eaten with vegetables and pulses. It is very good with lentils and beans and with cruciferous vegetables such as cabbage and cauliflower; many people believe it makes them more digestible. But not only vegetable dishes profit from asafetida; it is also good with fish, and goes into seafood curry. It is also mixed with ground meat for *kofta* (meatballs) and is important in a popular Kashmiri dish of lamb with yogurt sauce. All of India likes crispy asafetida-flavored *papadams*—those flat, fried, cracker-like rounds made from lentil flour. The spice is also an important ingredient in **chaat masala**. Many Hindus, particularly in Kashmir, eschew garlic and onions as inflaming the baser passions, and turn to asafetida instead. On the other hand, some Indian recipes go all out, and call for asafetida *and* onions and/or garlic.

Asafetida is sold both in powder form and in lumps. The lumps are generally supposed to be purer than the powder, and you might need to use more of the powder, but both forms are fine to work with. If you can find pearly or light gray lumps, grab them! Brown lumps are older or more adulterated or both, but they are still perfectly useable.

Many Indian housewives include a lump of asafetida with their stored spices to deter insects.

Asafetida is used in minute quantities; as little as one-fourth teaspoon of powder usually suffices in a dish for four persons. If you have the lump form, first rinse it under the tap and blot it with a paper towel, then chip off a piece about the size of a green pea, using a meat-tenderizing mallet or a hammer. Pulverize it in a small mortar, covering the top of the mortar with your hand to keep the bits from scattering; or put the asafetida between two sheets of waxed paper and hit it with the mallet. It will easily pulverize, and a pea-sized chip will yield approximately one-eighth teaspoon of powder. If the lump is good-quality asafetida, this should be as potent as twice as much of the powdered form.

Sprinkle the powdered asafetida over hot cooking oil and fry it for just a few seconds before the vegetables or other ingredients are added. Scatter the powder widely so that it will not clump together, and do not fry it too long or it may burn. If your recipe does not call for oil, scatter the powder over the hot liquid in a similar way.

From the first-century Roman cookery writer Apicius, we learn a trick that will make a lump of asafetida last indefinitely: put a large lump in a sealed jar with about a dozen pine nuts. The pine nuts will gradually absorb the flavor of the asafetida, and three or four of them, crushed, will substitute nicely for one-fourth teaspoon of the spice. Replace the nuts used with a few new ones, and seal the jar up again for future use.

If you cannot get asafetida, substitute one tablespoon freshly-grated white onion for one-fourth teaspoon asafetida powder, and treat it in the same way.

avocado leaf (*Persea americana*): Not every avocado tree produces leaves suitable for use as a flavoring; the best ones come from the hardy late-ripening Mexican breeds of avocados, rather than the West Indian or Guatemalan ones. The tree must be mature and grown in the right location (which may mean only in Mexico and the adjacent states of the United States). Mexicans say that the worse the fruit, the more flavorful the leaves are on a given tree. When you find a suitable source, the leaves can be dried and carefully stored for up to a year.

These long, shiny dark green leaves have an herbal **anise** taste, and are especially favored in Puebla and Oaxaca in Mexico. They are used in broths, stews, and *moles* (sauces), and are especially good with fish, chicken, and beans.

Lightly toast the whole avocado leaves, either slowly in a low oven or quickly on a grill. When they have cooled, you can store them in a sealed container, keeping them away from heat and light. The leaves may then be used whole, simmered in a liquid and removed, like a bay leaf; or they may be ground fine with the other seasonings in a mole, such as the famous thick, black, chocolate mole of Oaxaca.

Larger avocado leaves are also used as wrappers or to lay down a bed of flavor for foods to be steamed, such as chicken or fish. To intensify the flavor, place another layer of leaves on top of the food before steaming.

Hoja santa can substitute for avocado leaf, and vice versa. Or, for simmering in a liquid, you can tie up a **bay leaf** and one teaspoon of toasted anise seed in a piece of cheesecloth, in place of two avocado leaves.

B **balm/lemon balm/melissa** (*Melissa officinalis*): This herb is valued for its mild, sweet **lemon** scent. Use fresh leaves from your garden, because balm's delicate flavor does not survive drying very well. Keep the leaves whole until just before using, then tear them gently into pieces to release their flavor.

The tender new leaves are good in green salads and fruit salads, or mixed with other herbs in soups and omelets. Add a few balm leaves to stuffing for poultry or fish, and include them in an herbed rice pilaf. Most often these attractive crinkly, serrate leaves are added whole to iced tea, punches, cordials, and white wine; one or two leaves are sufficient for a bottle of wine. In Spain, they are also used to perfume milk for drinking. An infusion of balm makes a refreshing hot drink, with or without a little honey or sugar.

Substitute **lemon verbena** or lemon **thyme** in equal amounts.

basil (*Ocimum* species): Of the many species of basil, two have been cultivated for thousands of years: *O. basilicum* or sweet basil and *O. sanctum* or holy basil. Each of these species has numerous varieties or cultivars, some of which will be discussed below. These differ in their growth habits, size, and color, but what is most significant for the cook is their different proportions of essential oils, which determine their fragrance and flavor. Basil plants cross-pollinate readily and the seeds are extremely variable but, fortunately for the amateur herb gardener, most of the hybrids turn out to have an agreeable taste.

sweet basil (*Ocimum basilicum*): Sometimes called garden basil, common basil, or Genovese basil, sweet basil is the species most widely cultivated in the West, and is usually what is intended by recipes calling simply for basil. The flavor of any basil is dominant and the herb should be used sparingly: a couple of leaves are sufficient for almost every recipe. Keep the leaves whole as long as possible to preserve all the nuances of the flavor. Add basil at the end of cooking to keep it from turning dark.

"Basil should be torn—out of respect," opines one Italian culinary expert; and indeed tearing the leaves into pieces with your fingers, rather than chopping them with a knife, is usually the best way to deal with fresh basil, whenever you are using just a few leaves in a mesclun or other salad of tender greens, in a tomato sauce, or a fruit dessert. Exceptions to this rule occur when you are slicing a large number of leaves into a chiffonade or pounding them into a pesto.

Fresh basil leaves are excellent in a green salad. Try adding a basil leaf to a sandwich in place of—or in addition to—lettuce; this herb is especially good in a sandwich that includes cheese or tomato. Basil's great affinity for tomatoes is well known, somewhat obscuring the fact that its complex flavor also magnificently enriches fruit dishes. Tear a fresh leaf into tiny bits and scatter them over the bottom crust of a peach or apple pie before adding the filling; or serve an open tart or a bowl of mixed fruit with a decorative fresh basil leaf laid on top: its fragrance alone will flavor the food.

One very popular way to enjoy the flavor of basil is in an herbal vinegar (see Flavored Vinegars in the "Flavor Combinations" chapter). You can use either the fresh or dried herb for this purpose. A purple basil variety makes an especially beautiful, rose-colored vinegar with a gentle, flowery taste. Basil-flavored vinegar is an excellent choice for salad dressings, as well as for marinades. Try adding a few tablespoons of this elixir to a meaty vegetable stew.

Basil is important in Italian cuisine, and is closely associated with pesto, a rich sauce made of crushed basil leaves, **garlic**, pine nuts, olive oil, and a hard cheese, such as parmesan. Pesto on pasta is the classic welcome-home treat served to Genoese sailors. This flavorful sauce can actually be made with any herb, or with a combination of herbs, although it was basil pesto that first became popular in the United States and which remains the best-known today. Other combinations of oils, nuts, and cheeses may be used; sometimes **lemon** juice is added or the garlic is omitted. The very similar *pistou* from the south of France traditionally garnishes vegetable and bean soup.

The names *pesto* and *pistou* come from the pestle used to crush the basil leaves in a mortar. Today the mortar and pestle have generally given way to the food processor or blender; use the processor to make a pesto with more texture, or the blender for a purée.

Keep pesto in the refrigerator and, unless it has a large quantity of acid ingredients (such as parmesan or lemon juice), do not keep it longer than two or three weeks, depending on the temperature in your refrigerator. Never eat any pesto that seems abnormal for any reason! (See the discussion of botulism in the following chapter, under Flavored Oils.)

In the Middle East, from Greece through Turkey and down into the Arab world, basil is rarely used in the kitchen, although the herb is well known in this region. Girls are sometimes named Reyhan, meaning "basil." Basil's perfume is appreciated—the herb may be brought into church in Greece as a kind of incense, for example, and young Gulf Arab girls sometimes braid a few fragrant sprigs into their hair—but its taste is not a part of the cuisine.

Just next door, however, in Iran, fresh basil is eaten with relish and as a relish—that is, it is one of several fresh herbs set out on a platter, to be nibbled out of hand without any dressing, either to perk up the appetite at the beginning of a meal or to freshen the mouth at the end. A typical *sabzi khordan* might include basil, **marjoram**, **mint**, **tarragon**, and watercress along with radishes, spring onions, and chunks of feta cheese; it is always served with bread. This platter is especially pretty when purple basil is used. Basil is seldom dried in Iran, because it is not considered a cooking herb.

Basil is also popular in the lively cuisine of the Republic of Georgia, located on the eastern shore of the Black Sea. There, the leaves are cooked as well as eaten raw, in vegetable dishes and with beans, meats, poultry, and fish. Georgian cooks make use of dried basil in their distinctive spice mixture **khmeli-suneli**, which flavors soups, stews, bean dishes, and marinades for meat.

The numerous tiny glands on the surfaces of the leaves contain the complex essential oil that gives basil its pungent aroma and its distinctive taste. Some of the many basil cultivars produce different chemical mixes in their essential oils that result in flavor notes resembling **clove**, **lemon**, camphor, **cinnamon**, or **anise**. The proper cultivar for a dish depends on the flavor you prefer for the final effect. If you have a choice of basils, in the store or in the garden, sniff and taste a bit of leaf to see which seems more suitable. Keep in mind that sometimes the difference in their flavors is subtle; any sweet basil variety will stand in successfully for any other.

Given basil's great popularity, horticulturalists are engaged in virtually nonstop production of new cultivars. Here is a guide to some of the common sweet basil varieties used in cooking:

> *cinnamon basil:* The taste is just what the name implies; use these dark green, glossy leaves with sweeter dishes, especially fruit desserts. Tear up a couple of leaves and include them in a stuffing for pork chops.

> *bush basil:* An unusually high proportion of clove oil in this dwarf plant makes it a pungent herb.

French basil/fine-leaf basil: The tiny leaves of this variety have a delicate flavor, somewhat sweet, that is favored by cooks in the south of France.

lettuce-leaf basil: The extra-large crinkled leaves of this herb are mild and fruity.

licorice basil: The bright green leaves of this variety have a subtle anise flavor. They are particularly nice with fruit; float a few whole leaves in cold punch. They also taste very good with fish; lay a bed of these leaves under a whole fish before baking.

opal basil: Opal basil is a relatively new cultivar, and is the only wholly purple basil, without any green mottling; it makes a fine show in a salad, and will lend a rich pink color to vinegar for salads. Like the other purple basils with green or mottled green leaves, opal has fine flowery undertones.

piccolo basil: These narrow dark green leaves taste sweet and anisey, and are ideal for pesto.

spice basil: Technically a hybrid rather than a sweet basil variety, this broad-leaf basil gets its name from its unusually strong aroma and camphorous flavor. It is sometimes confused with holy basil (see below) because of its similar scent. Use sparingly.

lemon basil (*Ocimum americanum*): This natural hybrid has been elevated to species status because it breeds true from seed. Its light green, pointed leaves have a pronounced lemony fragrance and taste. Use it in marinades for poultry or tear up a couple of leaves and add to chicken or turkey stuffing.

holy basil/sacred basil/tulsi/tulasi (*Ocimum sanctum*): Since very ancient times, holy basil has been revered in India, and almost every Hindu household keeps a pot of basil in a place of honor. In spite of this—or perhaps because of the sanctity of the plant—basil is not a culinary herb in Indian cuisine.

The taste of holy basil is stronger than that of the sweet basils, and it is this species which is used in Thai cooking and in other Southeast Asian cuisines. Again, there are different cultivars with differing flavors:

bai horapa/red-stemmed basil: Tasting strongly of anise, this is the most popular basil in Southeast Asian cooking; it is called *rau que* in Vietnamese.

bai maenglak/Thai lemon basil: This variety has small leaves and a flavor that emphasizes lemon. Thais eat it raw in salads.

bai krapao/Thai holy basil: This basil has the strongest flavor of all, smelling and tasting intensely of cloves. In Thai dishes, it is always cooked.

Many American cooks are eagerly rooting and planting the Thai basils, which they buy in Oriental shops. This works fairly well, but be prepared for a change in flavor resulting from the change in climate and soil.

The holy basils used in Asia are often quite strong, so be very careful in using Asian basils in a European recipe that wants sweet basil. Taste to be sure!

In general, you can get good results using any sweet basil in place of any holy basil in Asian recipes. However, if you have a choice, the best sweet basil to use in Thai or other Asian dishes is licorice basil; this is especially good to replace *bai horapa*. For *bai krapao* substitute sweet basil together with a little **mint**.

If you have no basil at all, substituting a pinch of ground cloves works in some dishes.

bay leaf/laurel leaf/sweet bay (*Laurus nobilis*): Culinary bay leaves are also called laurel leaves, but do not assume that all laurel trees have edible leaves; some—for example, the mountain laurel—are quite poisonous! Similarly, there are other bay trees whose leaves are not suitable for culinary use, such as the West Indian bay tree, used to make bay rum aftershave.

Dried bay leaves do a good job of retaining their flavor and, in fact, the leaves are somewhat less bitter after drying than when they are fresh. Fresh leaves should be dried until they are brittle, then stored in an airtight container. One or two medium-sized leaves will flavor almost any family dish.

The standard way to use bay leaves is simply to throw them, whole, into the pot and let them simmer in the liquid; the leaves are generally removed from the dish just before serving. You might prefer to break the bay leaf into two or three pieces and put them in different parts of the pot, but do not break it up so fine that you can't retrieve all the pieces. Even after long cooking, the leaves remain too hard and sharp to be eaten; they can obstruct or even puncture the intestine!

Another way of using bay leaves is to create a flavorful bed for roasted or steamed meats and fish. Also, foods cooked *en papillote* are often seasoned with a bay leaf, which is included within the wrappings.

It is very important to note whether you are using imported Mediterranean bay leaves (the true *Laurus nobilis*) or the leaves of the California bay tree, *Umbellularia californica*, an unrelated species. The domestic leaves have a somewhat different flavor and are much more pungent. The imported leaves often come from Turkey, and are so labeled. California Bay Leaves are long, narrow, and pointed; a few may be joined together at the bottom on a stem. The Mediterranean bay leaves are much broader, and the point at the top is less sharp. Almost all recipes are written with the imported leaves in mind, so if you have the California variety, use half or even a third of the quantity the recipe calls for.

Bay leaves are utterly indispensable in classic French cooking and are an ingredient of the basic **bouquet garni**, used in soups, stews, and the fundamental stocks. *Court bouillon*, for poaching fish, is also flavored with bay leaf, as are many marinades. This bright, fresh, balsamic flavor is highly compatible with beef, and is vital in meaty, hearty dishes such as *pot-au-feu* and *daube provençal*.

Bay leaves are frequently included in standard **pickling spice** mixtures and are among the spices in shrimp and **crab boil**. They are good with any fish, especially the oily ones.

An interesting flavor note can be added to sweet custards by seething a bay leaf in the milk or cream (and then removing it) before adding the other ingredients. When storing dried figs or prunes, lay a couple of bay leaves among them in the sealed container to keep their flavor bright. For variety, drop a bay leaf in the water you are going to use for boiling carrots or potatoes.

Use twigs from the bay tree as skewers for shish kebab; they add much more flavor than the metal ones! These sticks are especially good with cubes of swordfish for a fish kebab; and you can get extra flavor by interspersing whole bay leaves between the pieces of fish on the skewer.

The bark of this tree was used as a flavoring in ancient Rome, and it still tastes good today. Finely grated bark gives an exciting, sharp bay flavor to dishes that do not have a liquid to simmer the leaves in. Spice cookies and dishes based on soft cheeses such as ricotta can be brightened by the addition of this grated bay bark; one-eighth teaspoon of freshly grated bark will suffice for a cup of ricotta or a cup of cookie dough.

The dark berries of *Laurus nobilis* are also edible, and were often called for in ancient Roman recipes under the name *bacca laureus*. Today, unless you have a thriving bay tree of your own, it is almost easier to get a baccalaureate from a university than to find these berries to cook with. **Juniper berries** or big black **peppercorns** are the best substitutes we've got. Incidentally, these bay berries have nothing whatever to do with bayberry, *Myrica cerifera*, whose aromatic, waxy fruits are used to make scented candles.

Indian and Asian books sometimes use "bay leaf" as the English name for certain leaves used in their cuisines, such as *tej pat* or **perilla**, but these are not accurate translations, and bay leaves should not be used. See **cassia** and **perilla** for further discussions of this point.

If you are out of bay leaves, throw two or three whole **cardamom** pods into the pot and let them simmer, removing them at the end of the cooking time; the cardamom will duplicate quite well that refreshing balsamic lift that bay leaves give to a dish.

borage (*Borago officinalis*): This tall herb is worth growing just for its beautiful blue, star-shaped flowers, which fade to pink as they age. These are very attractive to bees, a fact that earns the plant its nickname, "bee bread."

The light, fresh cucumber flavor of the flowers makes them as tasty as they are attractive in salads of all kinds, and they are particularly good in potato salads. Pick them from the plant just before using, because they wilt and shrivel rather quickly.

The flowers can be preserved very simply by candying. If possible, pick them on a bright, sunny day. The traditional method is to brush the flowers with a little well-beaten egg white, dip them in superfine sugar and let them dry. Due to the current concern about salmonella in raw eggs, however, you may prefer to use powdered egg white or gum arabic dissolved in warm water instead. With either method, if the humidity is low, the flowers will dry nicely spread out on waxed paper in the kitchen, but in humid weather, you should heat your oven just to the "warm" setting, turn it off, and set a tray of flowers on the bottom shelf with the door ajar. Traditionally, these sweetened borage blossoms are floated in cold fruit punches or applied to cakes and cookies as delicious decoration.

The leaves of borage are also useful in the kitchen. When small, they have the same cucumber flavor as the flowers, and they have a cooling effect, similar to that of **mint**, in cold summer drinks. The tender young leaves can be chopped and added to green salads and are sometimes used as a substitute for cucumber in salads, because some people find the herb easier to digest.

Young borage leaves have an affinity for yogurt, sour cream, and cream cheese. Float finely chopped leaves on yogurt soup, use them as a spud topping along with sour cream, or mix them with cream cheese as a spread for crackers or a dip for raw vegetables. Unfortunately, borage leaves do not keep their flavor well when dried.

Be warned: while the young borage leaves are a little hairy, the larger, older leaves are downright prickly. And the flavor of the larger leaves goes off a bit, tending to taste of overripe cucumber or even of fish oil! These are not recommended for eating raw, but when cooked they become a pleasant, mild-flavored, and nutritious green vegetable. Cut off the stems and cook them uncovered in an ample quantity of water to preserve their bright green color. Use a stainless or enamel pot; borage will discolor aluminum.

Because borage contains small amounts of toxic alkaloids, it is better eaten sparingly.

C **caraway** (*Carum carvi*): Brown, ribbed caraway seeds (actually the dried fruits of the plant) look a lot like **ajowan**, **anise** seeds, **cumin**, and **fennel** seeds. You can see some differences in appearance among these seasonings—ajowan is the smallest, fennel seeds are the largest, and caraway is the dark-est—although size and color can vary depending on the source or variety of your spice; there is, for example, a black cumin variety which, while not truly black, is darker than caraway. But it is the bold flavor of caraway that unmistakably distinguishes it from the other seasonings. Caraway has none of the licorice taste of anise or fennel seeds, nor the dusty sourness of cumin, nor the thymelike flavor of ajowan. Caraway seeds taste warm, sharp, and slightly bitter, and in many recipes they are nearly indispensable.

All of the above-mentioned culinary "seeds" improve in flavor with a light toasting in a dry skillet just before they are used; but with caraway, this is imperative. Caraway seeds are rather hard, and it can be unpleasant to bite down on them unless they have been softened by cooking. The seeds are sold whole, and generally used whole, but they can be toasted and ground for immediate use if desired.

The persistent flavor of caraway is extremely versatile, but it is perhaps in baked goods that we know it best today. The seeds are routinely worked into the dough of rye breads and pumpernickel. Old-fashioned English seed cakes were based on caraway seeds. Scones with caraway seeds make a nice variation; they are even better when something sweet is baked in too, such as currants, raisins, or candied orange peel. In fact, caraway is surprisingly good in sweet foods: try a little toasted, ground seed in spice cake or in cookies.

A couple of centuries ago, caraway "comfits" were popular in England. These were tiny sugar candies with a caraway seed in the center. (**Coriander** seeds and **anise** seeds were also sugar coated.) Comfits served both as sweets and as digestives, very much resem-bling the candied fennel seeds offered by many Indian restaurants today. Recipes sometimes listed caraway comfits among the ingre-dients for pastries; they were either baked in with the dough or sprinkled decoratively on top after baking. If you make your own comfits, be sure to toast the seeds first to soften them a little.

Caraway is a favorite spice in northern and central Europe, where it is indigenous. It is used in dishes made with pork, goose, and rich cuts of beef, and sprinkled over coleslaw, boiled cabbage, and sauer-kraut. Potatoes and egg noodles benefit from caraway, and cheeses are sometimes studded with caraway seeds, as well. A captivating

Ukrainian caraway soup is made by pouring boiling water, enough to cover, over a quantity of seeds (about one-half tablespoon per cup of soup), steeping for at least fifteen minutes and then straining this flavorful liquid into a clear vegetable broth. A little of this "caraway tea" is also excellent in gravies and sauces.

The German liqueur Kümmel is, as the name implies, an homage to this spice, the principal flavor in the brew. So popular is this drink that the verb *kümmeln*, whose primary meaning is "to season with caraway," has acquired the additional connotation of "tippling." Caraway is also the basic flavoring for Norwegian aquavit.

In a garden scene in *King Henry IV, Part 2*, Falstaff was invited to partake of a pippin "with a dish of caraways." We hope that the seeds had been toasted to intensify their taste and minimize their toughness, but in any case, as Shakespeare knew, the flavor combination of apples and caraway is a treat.

If you grow caraway in your garden, you can make use of the leaves and roots as well. Neither of these plant parts tastes like the seeds, but they are both mild and pleasant to eat. Use young, tender leaves in salads or chopped as a garnish, like **parsley**. The boiled roots are sweet and delicate, resembling parsnips or parsley roots.

Caraway seed and **dill** seed are similar enough in flavor that they can be substituted for each other in equal quantities.

cardamom (*Elettaria cardamomum*): This splendid spice offers an exhilarating aroma, a refreshing taste, and a warming sensation. Small, dark cardamom seeds grow in three rows within bright green pods measuring half an inch or less in length. The plant which produces them is a relative of **ginger**, but cardamom is so singular that it is the only species in its genus. Although it is indigenous to south India, this costly spice is now profitably grown in many parts of the world.

Cardamom is an extremely versatile spice. Its partisans claim that it goes well with every food and beverage, whether hot or cold, and the many different ways that cardamom is used in different parts of the world seem to attest to that claim.

Dubbed "Queen of Spices" in India (where black **pepper** is, of course, the king), cardamom has been important in Indian cuisine for thousands of years. Whole cardamom pods are tossed into curries and biryanis by the handful, to be set aside or eaten by the diners, as preferred. For other dishes, the small, dark seeds are removed from their pods and crushed; ground cardamom seasons tandoori chicken and enhances sweets such as *gajar ka halva* (carrot and milk pudding), rice pudding, and *gulab jamun* (fried balls of thick reduced milk, in **rose**-scented sugar syrup). A small cardamom pod usually simmers in *masala chai*, a hot, refreshing spiced tea always served with milk and sugar.

This spice is responsible for the distinctive flavor of Danish pastries, Swedish coffee cakes, Norwegian Christmas cakes, Finnish *puula*, and Icelandic *pönnukökur.* Scandinavian cooks lace cookies of all kinds with ground cardamom, and use it to season waffles, crispbreads, and many other baked goods. Cardamom adds the Scandinavian touch to Swedish meatballs and flavors *glögg*, the hot spiced wine popular at Christmas.

In the Arab countries of the Middle East, the outstanding symbol of generous hospitality is cardamom-flavored coffee. The use of this valuable spice is intended as an honor to the guest, and the larger, greener, more expensive cardamom pods are preferred whenever possible. Middle Eastern cooks also use the crushed seeds in large, date-filled pastries or in other dishes with fruits.

Cardamom's ability to blend with other foods makes it a feature of numerous spice mixtures around the world. It appears in Ethiopian **berbere**, Parsi **dhansak masala**, Moroccan **ras el hanout**, and Yemeni zhug, as well as the various curry pastes that are popular from India across Asia to Japan. Cardamom also blends into **garam masala** and **mulling spices**.

This agile spice also appears in American food products more often than its level of name recognition would suggest. Many processed meats—salami, sausages, frankfurters, and so on—profit from cardamom's pleasant, camphorous flavor. It is also used in soft drinks, liqueurs, and confections.

Don't neglect cardamom in your own seasoning repertory. For the very best results, buy whole pods, remove the seeds, and crush them just before use; three pods will yield enough seeds to produce approximately one-fourth teaspoon of ground cardamom. Commercially ground cardamom seed is also good when fresh, and you may prefer to use it.

Try a half-teaspoon of the powder in gingerbread and spice cakes. Taste what it can do to chocolate cake! It exalts fruits: add a little to apple or peach pie, fruit salads, and jams. And make it the principal flavoring in custard or ice cream.

Sometimes high-quality whole green cardamom pods are bleached white. This affectation in no way alters the flavor of the seeds but it makes the pods unattractive for using whole in a curry or stew.

Occasionally, **grains of Paradise** are called "cardamoms." These seeds resemble each other somewhat in appearance, and both plants are in the ginger family, but cardamom tastes warm and mellow, while grains of paradise are hot and sharp.

brown cardamom/large cardamom/Nepal cardamom (*Amomum subulatum*): This native of the eastern Himalayan regions produces large, hairy brown pods containing highly aromatic seeds similar to cardamom in flavor, but with a harsher and somewhat more camphorous or mentholic taste. Mostly consumed in India, they are used in the same way as green cardamom.

cassia (*Cinnamomum cassia, C. burmannii,* and a few other *Cinnamomum* species): These shrubby evergreen trees in the *Cinnamomum* genus are probably native to southern China and parts of southeast Asia, but today they are grown in many other regions of the world as well. The aromatic leaves, fruit, and bark of these trees have been popular spices throughout human history.

Be sure you do not confuse cassia, which belongs to the *Cinnamomum* genus, with the strong laxatives senna and wild senna, entirely unrelated plants in the *Cassia* genus.

cassia leaf/tej pat/tamalapatra/cinnamon leaf: The leaves and buds of the cassia tree, *Cinnamomum tamala,* sometimes called Indian cassia lignea, are also aromatic and have a cinnamon-like flavor. These leaves were used in the cooking of ancient Rome, where they were known as *malabathrum* or *folia malabathri.* Today they are a popular seasoning in northern India. Dried cassia leaves (also called cinnamon leaves) are sold under the names *tamalpatra* or *tej pat* throughout India and in Indian stores in America.

Cassia leaves are long, slender and smooth-edged, with a distinctive pattern of three long veins running the length of the leaf, from its stem to its pointed end. Look carefully, because these leaves are often mislabeled as "bay leaves." Many Indian cookbooks also use this mistranslation; generally, when an Indian recipe calls for a "bay leaf," what is really meant is *tej pat.*

A cassia leaf is *used* like a **bay leaf**, however—that is, it's simmered in hot liquid for a long period and removed before serving—but of course its light cinnamon flavor is very different from that of a leaf from a laurel tree. If you do not have a cassia leaf, substitute two or three cassia buds (see below) or a one-inch length of a stick of true **cinnamon**. Or you can use a sprig of **curry leaves**.

cassia bud: The dried unripe fruits of the cassia tree are also an important spice. They are produced commercially in China but, from 1949 to 1972, they were not available in the United States for political reasons. A cassia bud looks somewhat like a **clove** but smaller and darker; atop its stem the sepals of the spent cassia flower curl into a little cup to hold the small black fruit.

Cassia buds are usually used whole. They have a warm, mellow cinnamon flavor which is excellent with fruits. Stew them up with any fruit for compote, for fruit tarts and pies, or for fancy fruit punches. Mulled cider or wine is wonderful with this mellifluous spice. Cassia buds also have excellent preservative properties, and they are highly recommended as a pickling spice.

cassia bark/Chinese cinnamon/cinnamon: This spice comes from the inner bark of various trees of the *Cinnamomum* genus. It is sold in the United States and elsewhere as "cinnamon." See the **cinnamon** entry for a detailed discussion of this important spice.

celery (*Apium graveolens*): Sweet garden celery was developed only three or four centuries ago from the bitter Mediterranean herb known as smallage or wild celery. Since that time, the distinctive flavor of celery has become essential in standard French stocks and it is one of the aromatic vegetables that make up the classic **mirepoix**. With the bitterness gone, raw celery became a sensation in the

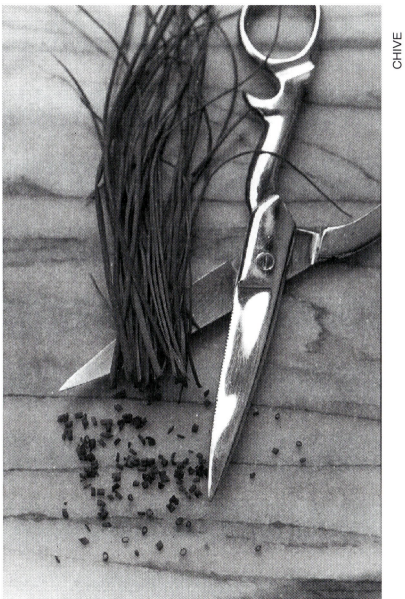

fashionable dining rooms of America and England in the last century, and tall cut-glass vases were designed to serve it in. Raw celery
stalks were so highly esteemed that they stole the show, and today
cooked celery dishes are unjustifiably neglected.

celery leaves are all too often trimmed from the stalks before
sale. But grab them when you see them, as they make a fresh,
flavorful addition to salads, soups, and stews. Dehydrated leaves
and small bits of stalk are sold by spice companies as "celery
flakes." They can be sprinkled into hot soups and stews early in the
preparation time, using half as much dried flakes as you would
fresh celery leaves. For uncooked dishes, the flakes must first be
reconstituted by soaking them in cold water for twenty to thirty
minutes; drain them well before using.

celery root, also called "celeriac," is the bulbous, knobbly root
of a variety of celery cultivated just for this part. Do not be put off
by the looks of this Cinderella vegetable: clad in its rough brown
peel, it is easily overlooked, but when it is properly dressed, the
results are magical. Not only can celery root can be treated like any
root vegetable, peeled and cut into chunks for stews or pot roasts, or
separately boiled, then mashed; it also makes a distinctive cold
salad. For salad, parboil slices of celery root in salt water, then
drain, cool, and dress them simply with lemon juice and olive oil; a
sprinkling of chopped dillweed (see the **dill** entry) is excellent on
this dish. When dealing with celery root, cut into the center to see if
there is a fibrous core; this should be removed.

celery seeds are the fruits of the wild celery plant, smallage.
Although they are tiny and lightweight, inside these minute seeds is
concentrated a great deal of the bright celery flavor. They are essential in **pickling spice** mixtures. Their intensity lends interest to
potato salad and other vegetable salads that might otherwise be
bland. Similarly, they are good in vegetable stews and fish stews, in
coleslaw, and in salad dressings. The brilliant celery flavor of the
seeds is also used to perk up breads and crackers, as well as tomato
juice and other vegetable juices and cocktails.

For best results, toast celery seeds before use (see Toasting Spices
in the "Culinary Practice" chapter), but only very briefly because
they are so small. Crush them slightly. The seeds are bitter and
potent, so use them sparingly and taste after each addition.

celery salt is a mixture of a lot of salt and a little celery seed. You get a flavor advantage if you make your own with your favorite **salt** and freshly toasted celery seed. Toast very lightly one tablespoon of celery seeds, remove from the pan immediately and cool. Then grind the seeds in a mortar with one-quarter cup salt. Use the same quantities of celery salt that you normally use of plain salt.

This favorite seasoned salt is sprinkled over deviled eggs, chicken salad, cooked carrots, or stewed tomatoes. Or use celery salt in place of regular salt in any recipe calling for celery seeds, to enhance the celery flavor.

chervil (*Anthriscus cerefolium*): A favorite in classical French cuisine, chervil is used much like **parsley**, but its flavor is sweeter and more delicate, with a suggestion of **anise**. Its aristocratic aroma and taste can easily be overwhelmed by other flavors, or driven off by too much heat. Cook fresh chervil leaves very little, if at all. They are best when added to a dish after it has been cooked, or toward the end of the cooking time. Do not hesitate to use them in quantity.

Fresh chervil leaves are added to green salads and to sauces such as béarnaise sauce. They are a standard ingredient of the classic combination **fines herbes** and a feature of *omelette aux fines herbes*. Whole chervil leaves are the traditional garnish for French sorrel soups, their warm, sweet flavor perfectly balancing the slightly sour sorrel.

Chervil's leaves are beautifully shaped, deeply cut into lacy patterns, almost like a fern. They are intriguing when interspersed among the florettes of a head of cauliflower. A small tuft of three to five leaves, often called a *pluche* in French, is fetching beside any vegetable.

Dried or frozen leaves can be used to supply that unmistakable chervil flavor to sauces, salad dressings, and to egg and mushroom dishes. You will need only about half as much chervil in this form as fresh. As always with leafy herbs, the texture suffers from freezing and drying, but the flavor will be adequate.

chiles/hot peppers/red peppers (*Capsicum annuum* and other *Capsicum* species): These fleshy, pungent New World fruits, which Columbus so infelicitously called the Native Americans' "pepper," have no botanical relation to the Old World's black **pepper**, but they do have the same power to heat the food and the body. (See the **pepper** entry for a discussion of black, white, and green peppercorns, and **pink peppercorn** for information on another unrelated plant.) Columbus noted in his log the Arawak Indians' word for this seasoning: *aji*. This name is still in use in parts of the West Indies and South America, and also refers to a hot spice mixture containing lots of ground chile.

After the Spaniards' arrival in the Americas, the chiles spread rapidly around the world. Under the influence of various climates and cuisines, the plant has hybridized into as many as 300 varieties, including the several colors of bell or sweet peppers, as well as **paprika**, pimentos, and all the hot chiles.

Today, the popularity of chiles in the United States is steadily increasing and a few years ago, for the first time, Americans spent more on salsa, which generally contains these piquant capsicums, than on our old standby, tomato ketchup.

The degree of pungency of a particular pepper is certainly vital, but remember that chiles differ in flavor as well. Both taste and hotness should be considered when selecting a chile for your dish. Mexican cuisine, which is very sophisticated in this regard, distinguishes among fresh, dried, and smoked chiles, with different names used for the same fruit in each condition, because flavor is altered when a chile is dried or smoked. Each form has its appropriate use; some sauces even call for several different types of chiles.

The taste of each chile is also affected by its degree of ripeness. To oversimplify the matter, a chile is green when it is unripe, and as it matures its color shades to yellow, then orange, and finally red. A ripe, red chile is sweeter and smoother-tasting, while a green chile has a brighter, fresher flavor.

People who like hot peppers usually *love* hot peppers. Some say a pepper isn't hot enough unless it makes you stomp your feet and holler out loud. Or, it isn't hot enough if it doesn't burn twice. Chile aficionados often act as if they were actually addicted to chiles, and there has been some speculation that they may be psychologically addicted to the high they get when the brain releases pain-numbing endorphins in response to the burn. If you're cooking for these "chileheads," be sure to give them some heat—even if you have to serve it separately in a hot-pepper sauce or a salsa.

The heat of these popular pods is supplied primarily by the tasteless, odorless chemical capsaicin. A reduced sensitivity to its effects can be acquired and maintained with repeated doses; or conversely, it can be lost by periods of abstinence. This means that the more you eat, the more you *can* eat; but if you take a long vacation from chiles, you will have to break yourself in all over again.

Most of the capsaicin resides in the pithy internal ribs of the pod, called "veins." It is also found on the seeds, which are attached to the ribs. These can be removed for a somewhat milder capsicum; peppers without their seeds and ribs are called *capones* in Mexico, that is, "castrated" chiles. Be sure to wash your hands thoroughly after dealing with chiles to avoid spreading the pungent chemicals to your eyes or other tender areas. With the hotter species, it is a good idea to wear gloves, because your fingertips are generally too sensitive to handle them. If you have neglected that last tip, you can console your irritated, tingling fingers with a paste of baking soda and water.

It is next to useless to reach for a glass of water to cool your mouth after eating chiles that are too hot for you. Water does not put out that fire. The best coolants are milk, yogurt, and other dairy products. Beer, bread, tortillas, and rice are helpful.

While black pepper makes itself felt immediately on the front of the tongue, many of the hot chiles seem to prefer a delayed action. They sneak down the back of your throat and, just as you are saying "Oh, this isn't so hot," they suddenly ignite. Other chiles, however, bite in different locations, attacking the lips or the tongue, as well as the back of the throat.

Hotness in a chile is sometimes measured in Scoville Heat Units, which range from 0 for a bell pepper to 1,000 for an Anaheim to between 4,000 and 10,000 for jalapeños and up to as many as 300,000 units for the habanero. As this system was originally devised, the pungency of a pepper was determined subjectively by the sensitive human tongue. A more accurate—and less painful—method is to use a modern machine which has been developed to analyze the pungent chemical compounds.

For those without a laboratory, the simplest method is just to rank the chiles relative to each other. Obviously, there is no way to cover all the many types used in cooking in the United States today, but a few of the better-known chiles are discussed here according to pungency, beginning with the hottest.

The habanero, the Scotch bonnet, and the datil pepper rate the top hot spot. This is serious. The habanero satisfies souls in the Yucatan, the Scotch bonnet is a favorite son in Jamaica, and the datil maintains traditional recipes in St. Augustine, Florida. Brave eaters from around the world endure the ferocious flare-up of these chiles for their sensuous perfume and their sweet flavor.

Unusually, these three are classified as varieties of *Capsicum chinense*, whereas most of the chiles we eat are *Capsicum annuum* species. *C. chinense* pods tend to be lantern-shaped, although the datil pepper is somewhat more pointy than the others. These pretty little devils are all generally preferred in yellow or orange stages of ripeness, and in fact the datil pepper rarely goes to red when ripe.

The Japanese favor a fearsome chile, the santaka, which claims to be among the world's hottest. This long, slender pod is generally ripened to bright red, then dried and powdered.

Tiny bird peppers, pequins, and chiltepins pack a big punch for their small size. These chiles are usually gathered in the wild, and most often used red-ripe in pepper sauces and for making pepper jelly, a particularly fine condiment for game. They may also be used dried as seasoning. The feisty pods attack immediately, burning lips, tongue, throat, and anywhere they touch. Birds love them.

Ground red pepper, or cayenne pepper, is a sharp, hot powder made from dried ripe chiles. The term "cayenne" is misleading, since any of several rather hot types may be used, not just the cayenne variety. The various brands of this seasoning differ in color and pungency, and a little experimentation will allow you to settle on your favorite. A teaspoonful of this piquant powder lifts many a dish out of blandness, and a fine sprinkle on top of pale dishes improves their looks dramatically. Ground red pepper is an ingredient in many different spice blends, particularly **chili powder**. (The variation in spelling is meant to indicate chili, the popular spicy red dish, rather than chile, the fruit of the capsicum.)

Dried pepper flakes serve the same purpose as cayenne pepper, but they give your dish a slightly more rustic or informal appearance. Some brands of pepper flakes have a particularly fine, distinctive taste.

Brazilians dote on a short, spiky little chile that burns with a high heat. Used in both its bright green and bright red stages, it goes into marinades and vinegars, and is indispensable in the sauce for Brazil's national dish, *feijoada*. Brazilians call it "malagueta pepper," but it is unrelated to **grains of Paradise**, which are often called by the same name. It seems likely that the Portuguese, who had earlier traded in the West African grains, brought the name to Brazil and applied it to the locally available substitute—a hot chile.

The Brazilian malagueta pepper is a variety of the species *Capsicum frutescens*. The only example of this species cultivated in North America is the tabasco pepper, famously used to make hot sauce.

The diminutive, pointed Thai pepper supplies plenty of heat to Southeast Asian cooking. The ubiquitous Thai green and red curry pastes derive their fresh or sweet tastes from the green and red stages of the chile, respectively.

Although it is not quite so ferocious as the Thai pepper, the fat little green serrano has a sharp bite. Its fine fresh flavor is well suited to Indian foods, as well as south-of-the-border dishes. The serrano is popularly used both raw and cooked.

The round, dark red-brown pod of the dried cascabel contains loose seeds that rattle when the pod is shaken, giving this chile its name of "little bell." It lends a nice nutty flavor—in addition to quite a lot of heat—to sauces, soups, and stews.

The fleshy bright green jalapeño is an American favorite. Its versatility makes it fun to use. Stuffed jalapeños—the fat pods slit and seeded, then filled with cheese, breaded, and deep-fried—are a popular Tex-Mex snack food. Slices of pickled jalapeño are an attractive topping for nachos, their color as intense as their flavor. And one of the culinary wonders of the wild West is the sweet, hot, candied jalapeño, a favorite in the Christmas season. Try a generous sprinkling of bright green bits of candied jalapeño over **chocolate** ice cream, and drive your senses wild!

If a jalapeño is allowed to ripen to red, and then smoke-dried, it becomes a wrinkled, dull-brown chipotle. Its sweet, smoky flavor and medium heat make this chile especially compatible with meats and with fruits. Relatively expensive, whole dried chipotle chiles are rare on the market, but they are almost always available packed in small cans with adobo or some other sauce. Think of these canned chipotles as a handy seasoning, providing a kick and a richness of flavor: remove just one from the can, chop or puree it, and add it to stews, sauces, vinaigrettes, or other salad dressings.

The poblano may be the most popular chile in Mexico, beating out some heavy competition. When fresh, this large, relatively mild, dark green pod just begs to be stuffed, and *chiles rellenos* may contain cheese, chicken, ground beef or pork, or rice. Poblanos are the chile of choice for *chiles en nogada,* stuffed peppers smothered in a creamy walnut sauce and popularly decorated with pomegranate seeds and flat-leaf **parsley** to replicate the colors of the Mexican flag.

When dried, the poblano becomes the dark-colored ancho, although it may also be called by the varietal names mulato or negro, which are brown or black, respectively. The dried pod is flat, wide and somewhat heart-shaped, and it adds superb flavor to many sauces and stews. The ancho is basic to the legendary *mole poblano,* that vibrant chocolatey sauce served with turkey.

The long, tapering pasilla is a dried chile, almost black and very flavorful. A little milder than the ancho, it is also very popular in Mexican moles.

Horticulturalists in the state of New Mexico have devoted themselves to the development of a not-too-hot but very tasty chile with endless uses. This New Mexican long chile—quite narrow and elongated—is used green or red, fresh or dried, and the dried pods are also flaked or ground into powder. The soups, stews, sauces, and salsas of Southwestern cuisine, and the decorative *ristras* of Southwestern decor, owe much of their popularity to this sweet, earthy, lively pepper.

The mildest of the chiles that still possess a degree of heat include the Anaheim, banana, cubanelle, and pepperoncini. Such sweet, meaty peppers are especially good when roasted.

Roasting a chile pod enhances its flavor, and also facilitates removing its tough skin. The chile should first be blackened and blistered over a flame or under a broiler, then sealed in a paper bag until it cools; the skin then easily flakes off. Use this treatment for any chile you intend to stuff.

When dealing with dried chiles, start by rinsing them well under a stream of cold water, preferably also scrubbing them with a soft vegetable brush: their wrinkled skins can collect an astonishing amount of fine dust. Scorch each chile lightly in a dry skillet over medium-high heat until it begins to puff up. Be careful not to use too high a heat; if you burn one, throw it out. After scorching, the chile is much more pliable. Cut off the stem, then slit the pod open and, if you wish to minimize the piquancy, remove the seeds and veins. Again, be careful not to rub your eyes as you do this, and remember that the color from the chiles will stain fabrics.

Next, put the chile in a small bowl, pour enough boiling water over it to cover, and let the chile soak. If you like, a more flavorful liquid may be substituted for the water in this step, such as hot vinegar, orange juice, or tomato juice. After about twenty to thirty minutes, the chile *and* its soaking liquid are buzzed in a blender. Add the sauce from your dish and blend all together into a smooth, flavorful purée. Then pour the purée through a sieve, straining out and discarding the fragments of chile skin and the seeds.

If you feel your completed chile dish is not hot enough, you can always add a few drops of hot sauce or a dash of ground red pepper. If it is too hot, however, reducing the pungency is harder. You can dilute the hotness by adding water, juice, or some other mild ingredient, or you can boil a few chunks of potato in your fiery liquid to extract some of the heat; the pieces of potato can be discarded or served to the chilehead in the crowd. Or you can serve some merciful cottage cheese or other dairy product, and warn your guests that they may find themselves wanting to stomp and holler out loud.

Store fresh and dried chiles in the refrigerator to keep their color bright and their flavor fresh, and to protect them from insects.

chive/onion chive (*Allium schoenoprasum*): Chives have the most delicate flavor of all the members of the onion family, and go well with other mild-flavored foods such as cream cheese, butter (see Compound Butters in the "Flavor Combinations" chapter), or the familiar baked potato with sour cream. (Note that many restaurants that serve spuds with toppings regularly substitute chopped green onion tops for chives, to the point that even some of their waiters believe that the sharp, oniony scallions are the same thing as exquisite little chives. This is a dastardly corruption of a great flavor team! Demand better for your potato.)

The hollow, tubular leaves of this plant make attractive bright green little rings when sliced at right angles (see Chopping Herbs in the "Culinary Practice" chapter). These Os bring *ahs* to salads, omelets, deviled eggs, soups, chicken, and almost any vegetable. Add them raw at the last minute to maximize their flavor.

The attractive lavender or pink chive flowers are edible raw, as in a salad, or lightly cooked, as folded into an omelet. Pull off the petals and sprinkle them over the food. Chive blossoms are also as beautiful as they are flavorful in an herbal vinegar. When you pick the blossoms from a chive plant, remove the entire stem with the flower because this stem is tough and unpleasant, and you don't want it to get mixed up with the tender leaves.

Try to find fresh chives, as they do not dry very well. Frozen chives retain their flavor, but not their texture. Perhaps the most successful preservation method is freeze drying. It is not necessary to soak freeze-dried chives; simply add them to your dish, and they will rehydrate almost immediately.

Chives are an important participant in classic French *haute cuisine*, and are considered one of the elegant **fines herbes.**

garlic chive/Chinese chive/oriental garlic (*Allium tuberosum*): This species of chive is distinguished by its flat leaves, white flowers, and garlicky flavor. The flavor is much milder than real garlic, making garlic chives excellent for salads and other foods where you want just a hint of the stinking rose. Many people feel that these chives make better garlic bread than garlic itself. Chop the herb very fine, mix it with melted butter and slather over slices of bread; serve warm from the oven. In general, garlic chives are remarkably good when heated, and are frequently added in quantity to stir-fries, or are put into pot-stickers. But be careful not to overcook them, which will diminish their lovely aroma.

Garlic chives are popular in China and Japan. Chinese dishes are often garnished with the white flower buds; the opened flowers are not used. The large, flat leaves are also found in Persian cuisine, both cooked with other herbs in herbed soups and stews or served raw in the popular pre- and postprandial plate of herbs.

chocolate (*Theobroma cacao*): The generic name *Theobroma*, meaning "food of the gods," was given to the cacao tree by the great eighteenth-century botanist, taxonomist, and chocolate-lover, Carolus Linnaeus, and it is the cacao tree which gives us that divine food, chocolate. This large-leafed equatorial evergreen sprouts its flowers—and thus its fruits—adventitiously from its trunk and larger branches. The fruits are hard, ovoid pods containing a sweet white pulp surrounding approximately forty seeds, or "beans." The pods range from six to twelve inches long, while the beans resemble large, somewhat irregularly shaped almonds.

Several varieties of the cacao tree are presently in cultivation: chocolate from the *criollo*, considered superior in flavor, is the rarest and most expensive; *forastero* chocolate, more abundant and cheaper, is probably the most commonly used today; *trinitario* and other hybrids of *criollo* and *forastero* yield chocolate of varying qualities.

It is a long way from cacao seeds to chocolate. When the pod is first harvested from the tree, the seeds do not taste or smell chocolatey at all. Pulp and seeds must first be fermented for several days, after which the beans are dried, traditionally in the sun, and then roasted. Finally, the outer shells of the beans are winnowed away, leaving the aromatic brown nibs, rich in fat and flavor. They also contain the stimulants caffeine and theobromine, as well as serotonin and other chemicals and nutrients that make chocolate a tonic for the human spirit.

The nibs are finely ground in heated mills into an unctuous dark cacao mass, also called chocolate "liquor." When this is simply cooled and molded into blocks, with no added ingredients, it becomes the intensely bitter unsweetened chocolate used for baking.

In the beginning, however, chocolate was a drink. In Mesoamerica, the cacao nibs were ground to a paste on a heated stone; sometimes seasonings were added. The luscious dark brown paste was then either used right away or formed into patties and stored. Added to water, the mixture was frothed (originally by pouring the liquid from a height, later by beating it), then drunk immediately. This is how chocolate has been made in America for at least a millennium and a half, and so it was made in Europe when it was introduced there by the Spanish. The wide range of flavorings that we know were used with this drink in the sixteenth century might serve as an inspiration to anyone interested in experimenting with chocolate today. Among the many New World additions to the chocolate drink were honey, **allspice**, **annatto**, **chiles**, Mexican mint **marigold**, and **vanilla**. Among the Old World spices that were tried soon after the beverage reached Europe were cane sugar, **anise** seed, **cinnamon**, black **pepper**, **lemon** peel, ambergris, and jasmine.

In Mexico today, commercially prepared tablets ready for making a chocolate drink are usually flavored with sugar, true cinnamon, and sometimes ground **almonds**. Water for chocolate is seething and full of energy, just about to break into a boil; this dramatic state has become an image in Mexico for someone who is on the verge of exploding with anger or pent-up desire. For each cup of this water, add about an ounce, or a little more, of the prepared chocolate, roughly chopped. The mix is then whirled in an electric blender until very frothy—or it may be frothed more laboriously with a traditional *molinillo*, a pretty carved stick with a fluted base, which is held between the palms and twirled vigorously.

In Europe today, hot chocolate is a favorite breakfast drink and is considered healthy for children. It is still made the traditional way, often in beautiful chocolate pots of silver or porcelain. The pots are traditionally shaped with a bulbous bottom, a straight handle at right angles to the pot, and a hole in the lid for the carved stick that is used to mix the drink before serving. Shavings of chocolate, variously flavored, are sold specifically for making hot chocolate, and aficionados believe that they get a smoother drink if they begin with the bits of chocolate in cold milk (or water) and bring it all the way to a full boil; this allows the starch naturally present in chocolate to thicken the mixture. The chocolate is not allowed to go on boiling, but is removed from the heat immediately.

Note that hot *cocoa* is a somewhat different drink from hot chocolate. Cocoa came relatively late in the history of this foodstuff. In 1828, the Dutch chemist C. J. van Houten patented a process for separating much of the fat from the chocolate liquor. This light-colored fat is cocoa butter, useful in making cosmetics, pharmaceuticals, and chocolate bars. After the fat is removed, the remaining dark solids are pulverized into cocoa powder. Van Houten further treated this powder with alkali to make it combine with water better; this treatment makes Dutch-process cocoa darker in color and, by neutralizing some of the acidity, alters the taste slightly to what is now commonly termed "Dutch chocolate."

Cocoa the drink, good hot or cold, is a mixture of cocoa powder, sugar, and milk, with a small pinch of salt. Like the hot chocolate drink, cocoa takes well to spicing. Try **cardamom**, **mint**, or a slim twist of organic **orange** peel. The list of possible seasonings is limited only by your own preferences. Prepared cocoa mixes are available to do most of the mixing for you, but be careful not to use them for baking.

To make molded chocolate bars, meant for eating rather than for drinking, the chocolate liquor is sweetened and flavored. Additional cocoa butter is then added to make the mass mold successfully. This step was first taken commercially in 1847 by Fry's Chocolate Company in Bristol, England, to be quickly imitated by many other chocolate manufacturers. Today's chocolate bars are made basically the same way. They generally also contain vanilla and virtually always incorporate the emulsifier lecithin, usually obtained from soybeans, to improve the texture and stabilize the mixture. The U.S. government requires a minimum of 15 percent by weight of cacao solids (also called cocoa solids) in the chocolate liquor for sweet chocolate, and 35 percent for bittersweet or semisweet chocolate. Any product with less than the required amount of cacao solids must be labeled "chocolate-flavored."

Sometimes a vegetable fat other than cocoa butter is added to the chocolate liquor, the usual choices being palm kernel, coconut, cotton seed or soya oils. The resulting "imitation chocolate" is a product with a higher melting point that is easier to work with and quicker to mold. It is especially used for dipping, making decorations, and molding into complex shapes. Also known as "compound chocolate," "confectioners' chocolate," or "summer coating," the various products on the market vary wildly, with differing proportions of cacao solids and cocoa butter, different choices for the added fats, and different amounts and types of sugar, milk, or fla-vorings. All this makes it imperative to read the labels carefully and to conduct a taste test of your own to determine which product you prefer.

The terms "sweet," "semisweet," and "bittersweet," when ap-plied to chocolate, are not well-defined indications of the amount of sugar added to the product. Each brand uses its own interpretation of how much sweetness these terms imply. You just have to taste.

Swiss pioneers in the chocolate industry gave us milk chocolate. Henri Nestlé worked at condensing milk, and in 1875, Daniel Peter successfully incorporated this new form of milk into his chocolate bars. As a general rule, European chocolate makers continue to use condensed, or concentrated, milk in making milk chocolate, while British and American companies use dry milk solids. United States regulations demand not less than 10 percent by weight of cacao solids for milk chocolate.

An important step in the evolution of fine eating chocolate was the development of the conching machine by Rudolf Lindt, who opened his chocolate factory in Switzerland in 1879. The machine is named for its original shell-like shape. During conching, the chocolate mixture is heated, stirred, and aerated for long periods, sometimes for several days; this produces a velvety smoothness and at the same time improves the flavor of the product.

The most recent milestone in the history of chocolate was the development of white chocolate, another Swiss innovation, which appeared shortly after World War I. This product essentially con-sists of cocoa butter, without any cacao solids, mixed with sugar, milk, and vanilla. Its virtue is its breathtakingly smooth and creamy consistency. Since this white substance contains none of the dark cacao solids from the chocolate liquor, it does not meet the U.S. government's standards of identity for sweet, semisweet, or even milk chocolate. Furthermore, most manufacturers of white choco-late add a small amount of some fat other than cocoa butter, primar-ily to achieve a brighter, whiter color. (Cocoa butter is naturally an off-white or ivory color, the exact shade depending on the variety of cacao beans used, as well as on their processing.) Although Euro-pean regulations allow a small amount of other fats in white choco-late, the United States prohibits the use of the term "white chocolate" for products that contain any fat or oil besides cocoa butter. At this time the nomenclature is in a bit of a mess, and these various white products are generally called something like "white confectionery coating," "white confection," or made-up names suggesting milk, snow, or pristine brilliance. Again, a personal taste test will guide you in choosing among the various offerings for your own white chocolate cheesecake, white chocolate fudge, white chocolate chip cookies, white chocolate ice cream, and other desserts or sweets.

Another important chocolate product is *couverture*. Although the French term might be translated as "covering" or "coating," it decidedly does not refer to the "imitation chocolate" confectionary or compound products described above. Rather, couverture identifies the very fine chocolate used by master chocolatiers and pastry chefs in producing exquisite candies, cakes, and pastries. High-quality products can make a huge difference in the flavor of your chocolate creations, and it is often worth using the best chocolate that you can afford. You are unlikely to find couverture in the supermarket, but it is available in specialty food shops or may be ordered directly from the manufacturer. Note that many of them refuse to ship their products during the hot summer months.

What makes fine chocolate fine? Several factors are involved. One of the key elements is the addition of cocoa butter to a chocolate mass. Couverture has a higher percentage of cocoa butter than most chocolates, ranging from 32 to 50 percent. You can actually *hear* all that cocoa butter in good chocolate: its presence produces a distinctive "snap" when a piece is broken. Pure cocoa butter has a low melting point—below body temperature—and it is this sublime fact that allows you to hold a piece of fine chocolate on your tongue, letting it melt luxuriantly in your mouth, while you experience all the nuances of its flavor and texture.

The amount of cacao solids present in fine chocolate is also a critical element. High-quality dark couverture chocolates may have as much as 70 percent cacao solids. Extended periods of conching reduce the size of the individual cacao-solid particles to under 20 microns, undetectable by eye or tongue.

The variety and provenance of those cacao solids is extremely important in determining the quality of fine chocolate, for soil, climate, cultivation, and processing have an enormous effect on the flavor of chocolate, as they do on all plant foods. Most chocolate manufacturers deal with these inevitable variations by adjusting their blends of cacao beans from different locations to maintain a consistent flavor, but one American manufacturer is turning this variability into a virtue by producing "vintage" chocolates, labeled according to the year and the plantation on which they were grown.

Another factor in the quality of chocolate is the roasting of the cacao beans. This is a delicate operation, with both time and temperature adjusted according to such factors as the variety of cacao, the moisture content, and the intended use.

Chocolate is sometimes said to be America's favorite flavor, although to be truly meaningful the statement really ought to specify just what food is being flavored (see **vanilla**). For some people, chocolate is a passion, and if you feel that way you will certainly enjoy a thorough exploration of the many products on the market. Depending on the type of beans chosen, the techniques of processing, and the unique recipe of the manufacturer, you can find chocolates with differing balances of sweet and bitter, with or without a sour component in their taste, and possessing nuances of almond, anise, caramel, coconut, coffee, orange, cherry, or vanilla flavors.

Happily, there are many, many ways that chocolate can be enjoyed. Besides those drinks and confections which have already been mentioned, recall that brownies and fudge were, until recently, reliably chocolate. Chocolate is the perfect flavor-variation for *crème brûlée*, *pôt de crème*, and other creamy desserts. Almost anything can be dipped in melted chocolate with surprising, happy results, from strawberries to nuts to candied ginger to pretzels and potato chips. Chocolate fondue, with chunks of fruit, pound cake, or cheese for dipping, is an easy, impressive dessert. Countless sauces, syrups, frostings, curls, shavings, sprinkles, cut-outs, and leaves serve as toppings and decorations. Chocolate can even be a plastic medium for edible sculptures.

Chocolate combines so well with coffee that the combination is named: "mocha" was originally the name of a port in Yemen from which fine coffee was exported. Many chefs include a little coffee in their chocolate concoctions as regularly as they add **vanilla**, using either a strong liquid brew or dry instant-coffee granules. Other popular partners with chocolate are cinnamon, mint, orange, and almost any kind of nut. Equally delicious, if less well-known, are combinations of chocolate and black pepper, chocolate and chiles, and chocolate and anise.

The good news about chocolate—even more good news!—is that this wonderful flavor need not be restricted to sweets. One of Mexico's most famous dishes is turkey with *mole poblano*, a dark, rich, smooth, savory sauce made of unsweetened chocolate, chile, garlic, tomato, sesame seed, nuts, anise seed, cinnamon, coriander seed, black pepper, and many other ingredients, which vary with each treasured version of the recipe. This sauce can be bought ready-made in cans, or as a basic paste ready to blend with a broth, or you can make it fabulously yourself from scratch. Whichever way you prefer to do it, *do it* for a taste experience not to be missed in this lifetime. This mole is also excellent with chicken, or poured over enchiladas, rice, or refried beans.

Chocolate's compatibility with meats does not stop with turkey and chicken: Remember the old trick of enriching a pot roast by adding a cup of coffee to the braising liquid? Well, a bit of baking chocolate in lieu of the coffee does the job even better. Use about an ounce of unsweetened chocolate with a three-pound roast for stunning results. Break the chocolate into small pieces and add to the liquid below the boil, stir well, then season as usual, with **salt**, **pepper**, **bay leaves**, **parsley**, **thyme**, ground **ginger**, or whatever you like in pot roast. And if you have any chocolate broth left over when the pot roast is gone, use it to make an extraordinary pot of beans.

Melting in a small piece of baking chocolate will make dark and mysterious tomato-based sauces for such savory dishes as spicy cocktail meatballs, caponata, or chili. And does anybody still make ketchup from scratch? An undertone of chocolate flavor in a home-made ketchup is devastatingly good. Chop the chocolate and add it after the sauce is done; keep the sauce below the boil to avoid burning the delicate chocolate.

storing chocolate and cocoa: When storing chocolate, you want to protect it from light, heat and moisture, and also to prevent it from absorbing foreign odors, which it is quite prone to do. There are horror stories of the confectioner with the remodeled pantry, whose hand-dipped candies smelled of varnish from the new shelves; and of the gift shoppe that sold chocolates that tasted of scented soaps.

Chocolate is best stored at room temperature, somewhere between 50° and 78°F. The cooler end of that broad range is particularly important for the finer chocolates, which contain much more cocoa butter. Rancidity is not a problem for this complex fat, but it does have a tendency to become unstable and separate out. Ideally, the humidity of the room will be under 50 percent; in no case should it exceed 65 percent. Wrap the chocolate tightly in foil—not in plastic, which may lend it a plastic aroma and therefore flavor. Properly stored, chocolate will keep for a year; dark chocolate will keep even longer, up to eighteen months or more. White chocolate, however, keeps less well, and should be used within a few months.

Improperly stored chocolate sometimes develops a "bloom"—a whitish or grayish film—when it is stored too warm and the cocoa butter in it melts and appears on the surface. Although unattractive, this chocolate has not spoiled; it can be redeemed by melting it over low heat and allowing the cocoa butter to mix back in. A more serious condition results when the chocolate has gotten wet, which often happens by condensation when chocolate is brought into a warm room from a cold refrigerator or freezer. This is sometimes called "sugar bloom," because the moisture may have dissolved some sugar out of the chocolate. It can make the affected parts grainy, and they should be reserved for some dish in which texture is not important. If, for some reason, you prefer to store chocolate in the refrigerator—perhaps your storage space is too hot or too humid—then wrap it well in foil, place it in an airtight container, and wrap that container in plastic. When you are ready to use the chocolate, let the entire package come to room temperature before opening it.

Cocoa powder, much of whose cocoa butter has been removed, is easier to store. Simply seal it tightly, and put it on a shelf at room temperature.

working with cocoa and chocolate: Cocoa powder meant for baking is easy to use: simply combine it with the other dry ingredients called for in the recipe. Because cocoa naturally contains a great deal of starch, you will have to make allowances for it if you wish to make a chocolate variation of a recipe that did not originally include that ingredient, either by reducing the amount of flour or by increasing the amount of liquid.

Many recipes call for melted chocolate. Make sure all utensils used are completely dry, as a few drops of water can cause chocolate to seize or "tighten" into stiff lumps. Chop up squares and large pieces to minimize the heat necessary for melting, because chocolate can easily burn or turn grainy when subjected to high temperatures. Never allow dark chocolate to reach temperatures above 120°F. Milk chocolate and white chocolate are especially liable to scorch, so take extra care and use even lower heat when dealing with them. Melted chocolate can deceptively retain its shape, so stir it well before adding more heat. If only a few small pieces remain to be melted, do not heat it any further, but simply stir until the bits are melted by the residual heat from the surrounding chocolate. Let the melted chocolate cool to nearly room temperature before using it in your recipe.

The safest methods for melting chocolate call for a microwave oven or a double boiler. In the microwave, place coarsely chopped chocolate in a suitable bowl. Heat it on medium power (50 percent) for thirty seconds. Stir and heat again for another half minute. Repeat as necessary. Microwave ovens vary, but two to three minutes should be sufficient total time in any case.

For the double boiler, begin by bringing a little water to a boil in the bottom pan; then turn down the heat. Put chocolate pieces into the upper pot and set the pot *over*, not *in*, the hot water. The water does not need to be boiling or even simmering to melt chocolate: merely hot water at about 100°F is sufficient, and it reduces the risk of steam condensing into the chocolate. When the chocolate has melted, dry the bottom of the upper pot with a dishtowel to be sure no drops of water accompany the chocolate when you pour it out.

Chocolate can also be melted over very low direct heat, although this is riskier. The method works best if your recipe calls for the chocolate to be melted in some liquid such as water, cream, milk, coffee, rum, or syrup. You must have sufficient liquid to keep the chocolate from stiffening; use at least one and one-half teaspoons of liquid per ounce of chocolate (generally one square). Add coarsely chopped chocolate to the cold liquid in a heavy saucepan, set it over low heat on the stove, and stir constantly.

If a little water does get into the chocolate and cause it to stiffen, it can be rescued by the addition of a little vegetable oil or solid vegetable shortening. Use about one teaspoon per ounce of chocolate. Butter and margarine, which contain water, will not work. Stir well until the mixture becomes smooth and supple again. Of course, this procedure alters the nature of your chocolate. It will be fine for cakes, cookies, and other baked goods, but not ideal for candies, custards, or sauces.

"Premelted chocolate" is available on the market to save you this step in cooking, but you have to sacrifice much in terms of quality for this convenience.

Chocolate cakes will richly reward your efforts to use a quality chocolate. These cakes mellow nicely if allowed to sit awhile; in fact, some cooks prefer to make them a day ahead, icing them on the day they are served. If you do hold the cake overnight, be sure to keep it in a cake box or wrap it carefully in foil, and set it in the refrigerator to keep it from drying out or absorbing other odors in the kitchen.

Because chocolate contains fat, in the form of cocoa butter, it will tend to deflate the beaten egg whites needed for chocolate meringues or chocolate soufflés. (See Leavenings in the "Culinary Practice" chapter). Gently *fold*—don't stir—the melted chocolate into the egg whites and do not overblend. For such dishes as these, the very best chocolate is not the ideal choice, since couverture is enhanced with so much additional cocoa butter.

tempering: When dealing with fine chocolate, you have to come to terms with the finicky nature of the cocoa-butter molecule. The process for dealing with this is called tempering, and it involves careful attention to temperature. Well-tempered chocolate has a smooth texture and a lustrous sheen. Its fluidity makes it easy to work with, and it contracts slightly as it sets, making it easy to remove from candy molds. Chocolate that is out of temper does not set well, and may be grainy or develop streaks of bloom as the unstable cocoa butter migrates to the surface.

Tempering is necessary for dipping or molding chocolate candies, for creating chocolate ribbons, leaves, and other decorative pieces, and for coatings of very fine chocolate. You do not need to temper chocolate used in baked goods, puddings, sauces, or for the centers of candies; only the chocolate covering the outside requires this treatment. It is also unnecessary to temper chocolate for a beverage: in that case, simply melt it. Compound chocolates and ordinary baking chocolate with a minimum cocoa-butter content do not need tempering.

All molded chocolate is in temper when it comes from the manufacturer, but you have to melt it to use it, and that puts it out of temper. As the chocolate sets up in the form you desire, you need to control the recrystalization of the cocoa butter to keep it at its most stable. Be sure you have a good chocolate thermometer. A candy thermometer will not do; it is set to measure very high heats, which would destroy your fine chocolate. A chocolate thermometer measures the range between about 70° and 130°F.

To temper fine chocolate, chop it into pieces and melt it in the top of a double boiler, as described above. (Many *chocolatiers* prefer to set a large metal bowl over a pot of hot water instead of using a double boiler, to eliminate any chance of steam from the bottom pot reaching the chocolate.) Watch carefully to see that the temperature of the chocolate does not exceed 115°F; aim for 113°F. When all the chocolate has melted, pour a quantity of it—from half to two-thirds of the potful—onto a cool, clean, perfectly dry marble slab. With a pastry scraper or a spatula with a broad flat edge, drag it around over the slab, working so as not to incorporate any bubbles, and keep it in constant motion so that it cools evenly. Scrape the mass back together, then spread it out again. Continue until its temperature is in the high 70s. Then add this worked chocolate back to the remaining melted chocolate in the bowl. Stir it to an even temperature, which should never exceed 90°F. (If it does get hotter than this, the chocolate is not spoiled—unless you somehow exceeded 120° and burned it—but the tempering must be started over. Set the chocolate aside until it has cooled before beginning again.) The desired temperature for the combined chocolate is between 86° and 90°F for bittersweet or semisweet chocolates, between 84° and 88°F for milk chocolate, and between 83° and 87°F for white chocolate. To maintain this temperature range, set the bowl over the hot water (off the stove), and check the temperature repeatedly as you work. If your chocolate gets too far out of its proper range, then retemper. For those who do a lot of work with chocolate, tempering machines are available that handle this problem extremely well, and will hold your chocolate in temper almost forever.

A number of chefs dispense with the marble slab and cool the melted chocolate by stirring in unmelted pieces. Some of them prefer to use finely grated chocolate for cooling, stirring in a spoonful at a time; others just add coarsely chopped bits; and still others move a single large lump around the melted chocolate until the temperature has fallen into the proper range, then remove the lump. With this method, chocolate does not hold its temper quite as long as with the classic marble-slab technique, but it is quite sufficient for chocolate creations that will be consumed by eager recipients within a few days.

substitutions: Milk chocolate works very differently from dark chocolate in most recipes, and they can seldom be substituted for each other—chocolate chips and coatings being the primary exceptions to this rule. Similarly, Dutch-process cocoa powder, treated with alkali, does not work the same way in all recipes as nonalkalized cocoa powder, because a baked product leavened with baking powder contains carefully balanced quantities of acid and alkali. (See Leavenings in the "Culinary Practice" chapter.) On the other hand, if you are simply making a cup of breakfast cocoa, then the type of powder chosen is only a matter of your own taste.

Although it is best not to substitute one form of chocolate for another, sometimes the following rough equivalences are adequate and/or necessary:

When your recipes calls for baking chocolate and all you have is cocoa, you can use three tablespoons of unsweetened cocoa powder plus one tablespoon of vegetable shortening in place of each square of unsweetened chocolate. But be sure to reduce the amount of flour in the recipe if you add cocoa; subtract one and one-half tablespoons of flour for each three tablespoons of cocoa added.

If you have unsweetened chocolate and your recipe calls for sweet or semisweet, substitute one-half ounce unsweetened plus one tablespoon sugar for each ounce of sweetened chocolate.

carob (*Ceratonia siliqua*): Also known as locust bean and St. John's bread, this seed pod comes from a tree related to the **tamarind**. The long carob pod contains a sweet, succulent pulp with a subtle fruity flavor, which long ago found a niche in Mediterranean

cuisines. The fresh pods are nibbled as a snack or made into a syrup to be incorporated into drinks or sweets. (If you get fresh carob pods, keep them moist and nibble-able by sealing them in a plastic bag and storing them in the refrigerator.)

In the past decade or so, carob has been popular with the food industry as a kind of substitute for chocolate. However, carob does not taste at all like chocolate, nor does it have chocolate's heady aroma. Its texture has an ineluctable dryness that will never mimic satiny chocolate. Furthermore, carob contains no caffeine nor theobromine—which, along with its low fat content, is why it is considered a health food. At most, we can concede that both carob and chocolate are brown.

You can use carob chips exactly like chocolate chips, but when replacing cocoa with carob powder—say, in brownies, cakes, or fudge—remember that carob is naturally sweeter than chocolate, and reduce the amount of sugar by about 25 percent. Since carob is also a milder flavoring than chocolate, you might also prefer to use 25 percent more carob than cocoa. Carob is enhanced by brown sugar, so consider replacing white sugar with brown whenever possible in cooking with carob. Finally, walnuts and carob have a particular affinity for each other, and these nuts make an excellent addition to carob-flavored baked goods.

The food industry also uses carob-seed gum, made from the endosperm of the hard flat seeds within the pods, to bind and stabilize such food products as salad dressings, ice cream, and pie fillings.

cinnamon: Our beloved spice cinnamon is derived from the inner barks of several different trees. Various members of the *Cinnamomum* genus, especially *C. cassia*, indigenous to Southeast Asia and perhaps China, are known as cassia cinnamon, while *Cinnamomum zeylanicum*, a native of the island of Ceylon (present-day Sri Lanka), produces Ceylon cinnamon, sometimes called true cinnamon. Historically, the two types have been regarded as two different spices; for example, the Bible specifies that both cinnamon and cassia were ingredients of a holy oil (Exodus 30: 23-25). In some countries such as the United Kingdom, spice dealers are required by law to distinguish between cassia and cinnamon, but in the United

States and many other countries, the bark of any tree of the *Cinnamomum* genus may legally be labeled "cinnamon."

Cassia cinnamon and Ceylon cinnamon resemble each other in appearance, and their tastes and aromas are similar, but there really is a flavor difference. Although they are used in much the same way, and each can serve as an excellent substitute for the other, cassia contains more essential oil and is more pungent, while true cinnamon is more delicate and complex. The trouble with the current system of nomenclature is that we in the United States have lost a useful refinement in flavoring our foods: some dishes are better with cassia, while other foods reach their peak with true cinnamon—and why not develop the savvy for that kind of seasoning precision?

cassia cinnamon/Chinese cinnamon/cassia (*Cinnamomum cassia, C. burmanii,* and other *Cinnamomum* species): Almost all of what is labeled "cinnamon" in the United States today is really cassia. American spice companies argue that the public prefers the more intense taste of the cassias, and they are correct in that cassia is indeed what most Americans are accustomed to. Intensely flavored cassia is the right choice for sweet rolls, sticky buns, and coffee cakes. It has a great affinity for apples and enhances apple butter, applesauce, mulled cider, and fruit-filled doughnuts, kolaches, and turnovers, as well as apple cobblers and good old American apple pie. Cassia is also best for cinnamon toast: choose a high-grade spice with a strong aroma, as it is the principal flavor of this homey dish.

As an indigenous plant in a wide area of Southeast Asia, cassia, rather than true cinnamon, is the authentic flavor for dishes of China, Korea, Vietnam, Indonesia, Malaysia, Burma, and other countries of that region. This is why cassia is sometimes called Chinese cinnamon. Cassia is one of the standard ingredients in Chinese **five-spice powder**.

In Chinese cooking, cassia is most often used with meats, especially pork. Sticks of it are added to stews and broths to simmer until the end of the cooking time, when they are removed. For dry cooking methods—roasting, barbecuing, frying, and stir-frying—the meat is first rubbed with a small amount of five-spice powder; the spices are then allowed a half hour or so to penetrate before the cooking begins.

Cassia also flavors the lovely marbled Chinese tea eggs (see **star anise**). Throughout the winter, Koreans enjoy substantial, satisfying chestnut sweets flavored with cassia. The Vietnamese clear-broth beef and noodle soups called *pho* are also seasoned with cassia cinnamon.

Saigon cassia, from the species *Cinnamomum loureirii*, was considered the best cassia in commerce until our supplies were cut off in the 1970s by the fall of Saigon and the subsequent U.S. ban on trade with Vietnam. Some spice dealers offer a blend of various powdered cassia barks that aims to approximate the fine, dark, high-oil-content Saigon cassia. Now that trade with Vietnam has resumed, Saigon cassia is available again. It is rather expensive, costing three or four times as much as ordinary cassia, but many chefs feel that its soft, warm, complex flavor is well worth the price.

Middle Eastern cooks also subscribe to cassia. They often toss a stick into the pot when making any soup containing lamb, or add half a teaspoon or so of the powder to dishes made with ground lamb. Cassia is also part of the traditional spicing for *kabsah*, a popular stew of Saudi Arabia, made with rice, meat, and vegetables.

When cassia is harvested, the fragrant inner bark is peeled from the slender branches of the shrub, rolled into sticks, and dried. The bark may be rather thick and is often rolled like a scroll from both edges toward the middle. Sometimes bits of the outer bark remain attached. These sticks—sold in the United States as "cinnamon sticks"— are usually cut into pieces three to four inches long.

The powdered form of the spice is best bought already ground because it is too difficult to grind it at home. The ground spice is appropriate for pies, cakes, and cookies of all kinds, for cinnamon sugar and for cinnamon-flavored butter (see Compound Butters in the "Flavor Combinations" chapter). Ground cassia cinnamon is included in pumpkin pie spice. Note, however, that powdered cassia loses its oomph quite quickly, so be sure to buy it in small amounts and store it tightly sealed.

For any dish that calls for ingredients simmered in a liquid, choose sticks over the ground spice; this is true for hot chocolate and cocoa, espresso and other coffees, mulled wine and cider, sugar syrups for fruits, and the hot milk or cream used in custards. The sticks give a far superior flavor, because they retain their essential oils better than the powder, and also because the powder never actually dissolves in a liquid: be it ground ever so fine, it will still taste floury on your tongue, and the tiny bits of powdered bark will eventually settle out and sink to the bottom of the container. Before you add the sticks to the pot, rough them up a bit by rubbing them very lightly over a grater. Pieces of the sticks are included in **pickling spice** mixes, both for flavor and as a potent preservative.

cinnamon/true cinnamon/Ceylon cinnamon (*Cinnamomum zeylanicum*): True cinnamon contains less essential oil than the cassia cinnamons, making it milder and more delicate in flavor, but it is unique in the *Cinnamomum* genus in that it also contains a certain amount of eugenol (also in clove oil) among its essential oils. Thus its flavor is also sweeter and more intricate than that of cassia cinnamon.

Where do you find true cinnamon? Probably not in mainstream American supermarkets, where virtually all the cinnamon sold is cassia cinnamon. But Ceylon cinnamon is generally found in Latin American markets and in Indian and Pakistani groceries, as well as in specialty spice shops. If you are unsure which spice you are dealing with, there are several tests to help you distinguish them: ground cassia is a darker and redder brown than ground Ceylon cinnamon, which is more a medium tan color. In sticks, cassia bark is thicker than true cinnamon bark, and a stick of cassia is usually formed of a single piece rolled inward from both sides like a scroll, while multiple thin strips of true cinnamon bark are frequently rolled up together into layered "quills." (True cinnamon is also sold in thin, broken bits, which are less attractive but entirely usable and very flavorful.) Unless the bark you are dealing with is old and dry, you can usually determine which cinnamon type you have by a taste test: chew a bit of bark thoroughly. If you can detect a slightly slippery or gelatinous substance in your mouth, then you have true cinnamon; otherwise, it is probably cassia.

True cinnamon seems to be native to Sri Lanka, and the best cinnamon still grows there today. India, Madagascar, and the Seychelles also produce high-quality true cinnamon, and that spice is preferred to cassia cinnamon in the cuisines of those countries.

True cinnamon is also preferred in Latin America, North Africa, and in parts of Europe, where it is generally confined to sweets. The ground bark of true cinnamon is considered better than cassia for fine pastries and creams, but European opinion seems to be divided

on whether true cinnamon or cassia cinnamon is best with chocolate. (But *both* are good with chocolate: try adding a couple of
teaspoons of either kind of cinnamon next time you bake a chocolate cake.)

True cinnamon is much loved in Mexico. It is routinely included
in chocolate, and often flavors coffee or is brewed into a tea. Mexican and Spanish *horchata*—a chalky, refreshing drink based on
ground tiger nuts (or rice, melon seeds, or almonds) with sugar,
water, and ice—is popularly flavored with a generous measure of
cinnamon.

Cinnamon is extremely important in Indian cuisine. It is almost
inevitable in elaborate curries and elegant biryanis, but also appears
in simple rice or lentil dishes. Often small pieces of true cinnamon
bark are used, which become tender enough simply to eat with the
rest of the dish. Cinnamon is an ingredient of the common mixture
garam masala, which is most often used in powdered form.

True cinnamon is often used in the richly spiced Moroccan cuisine.
It seasons the filling for the famous pigeon pie *bstila* and, mixed with
powdered sugar, also decorates the pastry on the finished product.

clove (*Syzygium aromaticum*): This venerable spice is actually the
dried bud of a flower, with its sepals and a bit of its stem. It is the
bloom of a tropical evergreen tree in the myrtle family. Pink when
picked, the bud dries to brown, with a round, tan head—the unopened flower petals—held like a jewel in a darker-colored setting.
With their wide heads and pointed stems, cloves are shaped somewhat like a nail, and we derive our name for this exotic spice from
the Latin word for that prosaic little object.

It is merely an unfortunate coincidence that we speak of a
"clove" of garlic or shallot: that word comes from the English
"cleave," meaning "to separate by splitting," and is applied to sepa-
rated parts of these bulbs. Less coincidental is the name of the clove
pink, *Dianthus caryophyllus*, the ancestor of the cultivated carna-
tion, for this flower has a scent very similar to that of the spice. In
German, *Nelke* is the word for the flower, while *Gewürznelke* is the
name of the spice. And in Turkish the same word, *karanfil*, is used
for both flower and spice. The clove pink and clove-scented carna-

tions have been used to lend a spicy aroma to wines, ales, and vinegars.

If the buds are left on the clove tree, they blossom into a pink flower and, after pollination, form a purple fruit resembling a small dark olive. Very rarely, this fruit is dried or preserved in sugar and sold under the name "mother of clove"; it tastes like a very mild clove.

Cloves are rich in the valuable essential oil eugenol, or oil of cloves, which is anesthetic and antiseptic as well as aromatic. (Eugenol is found in other plants as well; for example, it is distilled from the leaves of the **allspice** tree.) This oil is used in many processed foods, as well as in medicines, dentistry, and perfumery; it has also been employed in the manufacture of vanillin, which is used as a substitute for **vanilla**.

An old Chinese tradition says that before anyone approached an emperor of the Han dynasty, the subject was required to put a clove in his mouth so that his breath would be sweet. The technique still works today: chewing a whole clove will take away the reek of **garlic** or other pungent food from the breath. A good fresh clove will feel very hot at first, after which its anesthetic properties numb the mouth for a while.

Cloves blend well with other spices, but because of their potent fragrance and flavor, they should be used sparingly to avoid overpowering the other seasonings. They are associated with the "sweet spices," such as **allspice**, **cinnamon**, **mace**, and **nutmeg**, which commonly season spice cakes, pumpkin pies, Dutch *speculaas*, and gingerbreads. A popular Middle Eastern butter cookie, a feature of many festive occasions, sports a whole clove pressed into its center. The spice provides a perfect flavor accent for the sweet, buttery cookie, and the baked clove is milder and easier to chew than cloves in the raw.

Cloves are essential in **garam masala**, and appear in many **curry powders**. They are counted in **quatre épices** and in **five-spice powder**. Cloves are an important **pickling spice**, and many pickle and chutney recipes rely on them. Cloves also marry well with fruit dishes of all kinds, including mulled wine and cider. Baked apples benefit from this spice, and a little ground clove always flavors mincemeat pie.

Cloves also add richness to savory foods. One or two whole cloves are always included in classic French stocks; they are usually stuck into an onion or other vegetable so that they can easily be retrieved.

Clove-studded baked hams are famous. Score the fat and insert a large plump whole clove at the center of each diamond section for a beautiful effect and a complementary flavor. Be careful when handling the cloves: the heads are surprisingly fragile and are easily broken with the fingers.

Ground cloves or clove oil enrich a good many sausages, cold cuts, and other processed meats. Chicken dishes profit from an unobtrusive pinch of ground cloves; even fried chicken is improved by a hint of ground cloves in the breading. A single clove dropped into a beef stew or *pot-au-feu* is significant but unobtrusive. The popular German spiced beef sauerbraten is seasoned with cloves, among other spices.

Try a pinch of cloves in a pot of baked beans. A light dusting of powdered cloves works wonders for sweet potatoes, squash, or pumpkin. One famous old American recipe for tomato soup calls for a fistful of whole cloves—about three cloves per cup of soup, to be a bit more precise; these simmer with the tomatoes and other ingredients in a little beef broth and are strained out before serving.

Cloves require careful storage. Use your best airtight container, and return it quickly to its cool, dry, dark place. Ground cloves are particularly liable to lose their volatile oils, and also suffer in high humidity, which can cause the powder to form clumps. Since whole cloves are so difficult to grind at home, the best plan is to buy the commercially ground spice in very small quantities. It is tempting to obtain your powdered cloves from whole ones by simply crushing the heads with the fingertips, reserving the hard stems for another use—and you might certainly resort to this in an emergency—but this is not ideal as a regular practice, because the two parts of the "nail" actually have rather different flavors, with each making a contribution to the overall effect of the spice.

In a pinch, allspice can be used in place of cloves, and vice versa.

Although clove trees are native to the spice islands of the province of Maluku, the spice appears surprisingly seldom in Indonesian cuisine. When it is used in cooking, it is usually in curries, a reference to the Indian origin of these dishes. Nevertheless, cloves *are* used in quantity in Indonesia, to the point that the country must sometimes import supplements to its domestic supply: ground cloves are mixed with tobacco to make those arrestingly aromatic, crackling *kretek* cigarettes, so wildly popular in Indonesia and elsewhere. About half the world's production of cloves goes up in smoke every year!

coconut (*Cocos nucifera*): In coastal regions of the tropics all around the globe, the versatile coconut palm offers its trunk for structures and utensils, its fronds for thatch, its fiber for ropes and mats, and its seed and sap for a variety of nutritious foods. For many people, this is truly the tree of life. The fibrous fruit contains the largest known edible seed—the coconut.

Usually the coconut is offered for sale with its thick, green, fibrous husk removed. Select your hairy brown coconut carefully, checking that no cracks or holes are visible. Shake it to be sure that there is still some liquid inside: you can hear it sloshing if there is. This will tell you something about its age, because the older the coconut, the drier and harder it is. Store the unopened nut in the refrigerator; it should be used within a month.

To deal with a coconut, get out your arsenal! The peoples of the Caribbean master a nut neatly with a machete, and natives of the South Pacific can handle it with just a sharp stick planted firmly in the ground; but a hammer, screwdriver (preferably Phillips), short knife, and vegetable peeler are recommended for the inexperienced. An oven is also a useful tool in this process.

coconut water: Open the coconut's "eyes" with the screwdriver by placing it in position and tapping it lightly with the hammer. Turn the nut over and drain the liquid into a bowl. This is coconut water or juice, not coconut milk (see below). The water is a favorite beverage around the globe, and as nutritious as it is delicious. In the tropics, young men skillfully make the dizzying climb up the trunk of a tall coconut palm to cut down the nuts for the pleasure of drinking the pure, refreshing liquid inside. Coconut water is usually drunk immediately as a reward for the effort of obtaining it, but if you wish you may store yours in the refrigerator for a day or two.

Freeze it if you wish to keep it longer. This refreshing drink can also be bought in cans, usually sweetened and diluted with a little water. Chill the cans, or serve the juice over ice. When the liquid comes from immature green coconuts, called "jelly nuts," bits of the tender, gelatinous, unripe coconut meat are often included.

Coconut water, especially the more concentrated juice from older nuts, is used as a braising liquid to flavor and tenderize meats, and occasionally to enrich the liquid in a stew.

coconut meat: After you have drained the nut, heat it in a medium (350°F) oven for five to ten minutes to make it easier to open and to remove the shell. Wrap the hot nut in a dishtowel and set it on a hard surface. Whack it sharply a few times with your hammer until it breaks into two or more large pieces—the towel will keep them from scattering. The meat can now be pried off the shell with a short, sharp knife.

The white meat has a thin brown skin attached, which may be left on for some dishes. If you want snow-white coconut meat, you can easily remove this skin with a paring knife or vegetable peeler.

Refrigerate fresh coconut meat until you are ready to use it, and freeze it if you intend to keep it longer than a week. Do not eat fresh coconut meat if its smell, texture, or taste lead you to suspect that it has turned rancid.

Do you like potato chips? Corn chips? Try coconut chips! In the Caribbean, this easy-to-make, satisfying snack food is a favorite with drinks. With a vegetable peeler or a food processor, slice a piece of coconut meat, with or without the brown skin, into very thin strips, making them as long as possible. Bake the strips in a single layer on a cookie sheet in a low oven (about 300°F) for about half an hour. Watch them closely, and remove them from the oven when they have turned a deep golden brown. Sprinkle them with salt to taste while they are still hot, and serve.

You can shred fresh coconut meat with a grater or in a food processor. Or you can buy it in packets ready-flaked or grated, either finely or coarsely. North American cooks are most familiar with shreds of desiccated coconut with a thick sugar coating, but a far superior flavor is obtained from the unsweetened dried coconut that is reliably available in Indian and most Asian shops. The latter also allows you control over the sweetness of your dish.

Shredded or grated coconut is well known as a topping for cakes, pies, and other sweets. You can vary the look occasionally with golden-brown toasted coconut. Just heat unsweetened shreds in a dry skillet over medium heat until lightly browned. Grated coconut is also a popular means of improving macaroons. Mixed with a generous quantity of **orange** segments, shredded coconut makes ambrosia, a Christmas treat that can serve as salad or dessert, depending upon how much sugar you add. Try substituting grated coconut for the bread crumbs in your best meatball recipe for a tasty variation on an old favorite. These meatballs are even better if you add a little **coriander** seed, toasted and ground; coriander is a spice that marries most happily with the coconut flavor.

Keep dried coconut in a tightly sealed container, and store in a cool, dark, dry cupboard. It should last for many months.

Many Oriental shops offer fresh-grated coconut frozen in packages. This is very good, but it is not dry and should be used in dishes that can handle a bit of liquid, such as coconut chutney or a coconut sambal.

coconut milk and coconut cream: Shredded coconut is also the source of coconut milk. This rich, milky-white liquid is obtained by pouring boiling water over the grated flesh of a fresh nut and letting it soak until cool. Use equal volumes of grated nut meat and water. Strain, pressing down on the coconut in the sieve to get as much of the milk as possible. Repeating this process with the grated coconut left in the sieve yields a thinner milk, and the two pressings are generally mixed together for a medium-rich milk. If necessary, dried *unsweetened* coconut can be used instead of fresh; soak it several hours or overnight before straining out the coconut.

Coconut milk of excellent quality is sold in cans, and saves both time and trouble in the kitchen. You may even find "lite" coconut milk, with a lower fat content than normal. Check all labels to be sure you are not getting added sugar, water, preservatives, or other ingredients you may not want in your recipe.

Coconut milk is often called for in Southeast Asian recipes. It can mellow and tie together the full range of Southeast Asian seasonings. **Chiles**, **ginger**, **lemon grass**, kaffir **lime** leaves, **screwpine**, and **tamarind**, for example, are all happy partners with coconut flavor.

Indian spices such as **coriander** leaf, **cumin**, **curry leaf**, **mustard** seed, and **turmeric** blend beautifully with coconut milk in curries, mulligatawny soup, *rhogan gosht,* and countless other dishes.

Caribbean cooks use coconut with all sorts of tropical fruits—bananas, limes, mangoes, papaya, pineapples, and so on—and with vegetables of the region as well. A specialty of the island of Antigua is a mix of grated coconut and grated sweet potato combined with a bit of flour and seasonings such as sugar, **nutmeg**, and **vanilla**. Handfuls of the stiff mixture are wrapped in a banana leaf or a sea-grape leaf, and boiled. This dumpling-like *doucana* is traditionally served with salt fish.

The coconut is important in the South Pacific islands, too. There it is often combined with breadfruit and taro, and adds excellent flavor to all kinds of fish dishes. Coconut puddings thickened with cassava root (tapioca) are enhanced with tropical fruits.

But coconut also goes well with an astonishingly broad range of flavors from every part of the world. Try braising chicken breasts in a liquid comprising tomato sauce, coconut milk, and chicken broth. Or tame a Cajun jambalaya with a large dollop of coconut milk in the broth. Coconut milk is a boon for almost any eggplant recipe. Rice cooked in diluted coconut milk is an extra-flavorful side dish for almost any meal.

Coconut milk separates when it stands for a while: stir to blend, or pour off the thicker part on top to have coconut cream. As the term suggests, coconut cream is sweeter, thicker, and contains more fat than coconut milk. This, too, is available in cans and in blocks resembling hard margarine. To use, just cut off a tablespoon from the block, and let it melt into stews or curries for an extra dose of flavor. If you want liquid coconut cream as the cooking liquid of a dish, stir two and one-half tablespoons from the block into a cup of hot water. Coconut cream is also found dehydrated and powdered, sold in boxes or foil packets with intriguing recipes and serving suggestions printed on them. Be sure to read the labels of all cans or packets, as coconut cream is often intended for desserts and may be presweetened. "Crème de coco" often indicates a mixture of coconut milk, coconut oil, and sugar, meant for piña coladas and other tropical cocktails. You may find you prefer these drinks made from scratch with coconut cream, so that you can determine the amount and type of sugar added.

You can use coconut milk and coconut cream almost anywhere you would use the corresponding dairy products. An emphatic coconut custard can be made by substituting coconut milk for half the milk called for in the recipe; a pinch of salt and a dash of vanilla extract are needed for full flavor harmony. Top the baked custard with a sprinkling of toasted coconut flakes. You can also make your custard with all coconut milk, if you prefer, but about 50 percent more beaten egg will be needed to make the custard firm. This is an opportunity to eliminate dairy food for vegans, for cooks keeping kosher, or for people who are lactose intolerant. A thin smear of coconut oil, instead of butter, can even be used to grease the custard cups, but be careful not to use too much, because of its strong flavor. You may find the resulting dessert has an unexpected pale green tinge, but this can be concealed with a toasted coconut topping.

Refrigerate coconut milk or cream. If you wish to keep it longer than about three days, it should be frozen. Once defrosted, the liquid may seem a little grainy, but the flavor will still be good.

coconut oil: Dried coconut meat is the valuable tropical product copra, from which coconut oil is expressed. (Small amounts of oil can also be boiled out of fresh coconut milk.) This oil has a wide range of uses in the cosmetics and food industries. The ability of coconut flavor to blend beautifully with other ingredients makes it a favorite in processed foods. Coconut oil is the basis for dairy substitutes such as coffee creamer and whipped toppings. It is still used to make margarine, although in the United States a less saturated fat is now generally preferred for this purpose.

In much of the tropics, coconut oil is the most popular cooking oil, and is needed to give the authentic flavor to Southeast Asian and coastal Indian dishes. The oil is strong flavored and strong smelling; it smokes at relatively low temperatures but still works well for frying and sautéing.

Coconut oil solidifies at room temperature, between 74° and 75°F. When chilled—and it should be kept refrigerated—it is white and quite hard.

coconut gel: Strips or balls of gelatinous, translucent coconut meat are sold packed in sugar syrup. Known as *macapuno* in the

Philippines, this is obtained from a coconut sport whose meat does not become hard as the nuts mature. Sweet, mild-flavored, chewy morsels of coconut gel can be added to tropical fruit salad or mixed into custards and ice cream.

A similar product, available in cans, is young coconut meat in syrup. The flesh of the young coconut is more tender than the coconut gel, but it also has less flavor. It can be used in the same way as the gel, and it also makes a very nice cooling side dish to serve with spicy-hot foods.

coconut sugar: The sap from unopened flower stalks of the coconut palm and other palms is tapped and boiled down into a brown, grainy, unrefined sugar with a distinctive winey taste. This is one of the many unrefined sugars called "jaggery," generally available at Indian and Southeast Asian food shops. If you cannot find jaggery, substitute dark brown cane or beet sugar, or a less refined cane sugar such as turbinado or demerara.

coconut vinegar: A white, low-acid vinegar, made from fermented coconut water, is popular in the Philippines, where a subtle sourness is enjoyed in many dishes. The Philippine national dish, *adobo,* is based on chicken and pork marinated in coconut or other palm vinegar, which is often sold in tall bottles in Asian food stores.

coquito/coquito nut/miniature coconut/baby coconut (*Jubaea chilensis*): Appreciated as a fun snack food, this small fruit of a Chilean palm tree looks just like a tiny coconut, about an inch in diameter. And it tastes like a coconut, too. Pop a whole one into your mouth and chew—and chew and chew—it up, brown skin and all.

Small pieces of coquitos in biscotti, cookies, or other baked goods simultaneously supply the flavor of coconuts and the texture of nuts. Chop them up in a nut chopper or food processor if possible, because holding those roly-polies while you're trying to cut them with a knife can be hazardous.

Like the coconut, the coquito nut should be stored in the refrigerator.

coriander (*Coriandrum sativum*): This clever little herb supplies us with two important seasonings—the small round fruit (called

"seed") of the plant and its leaf. Even though they come from the same plant, they have entirely different flavors.

coriander seed: It is only as coriander seeds ripen and dry to a tan color that they develop their distinctive sweet, citrus taste and light, balsamic aroma. Many cooks prefer to buy whole seeds and give them a quick toasting (see Toasting Spices in the "Culinary Practice" chapter) before grinding. The seeds crush easily, so you can avoid commercially ground coriander, which loses its volatile essences quite rapidly.

For centuries, coriander seeds have been used in Europe and Great Britain to flavor spice cakes, gingerbread, and other sweet breads. Whole seeds have been covered with sugar and eaten as "comfits," and the sugar-coated seeds have been incorporated into cakes. Try the old-fashioned trick of sprinkling in a teaspoon of ground coriander when you make an apple pie!

Amelia Simmons, author of America's first cookbook, published in 1796, offered recipes for coriander-flavored cookies, and it's an idea well worth reviving today. Start with any basic cookie recipe, leave out the vanilla or any other flavorings, and add about one tablespoon of ground coriander seed for each three cups of flour.

Today, coriander seed often goes into sausages. It is good with beans and in pea soup. It is also an ingredient of shrimp or **crab boil** seasoning and **pickling spice**. This spice is particularly important in curries. Southeast Asian cooks make regular use of the seeds. Coriander also flavors Ethiopian **berbere** and Georgian **khmeli-suneli**. Egyptian cooks rely on *taklia*, a basic seasoning of crushed coriander seed fried with as much garlic as you like, which is used in lentil dishes and with all kinds of vegetables, including the national favorite *melukhiya* (*Corchorus olitorius*), a tasty, mucilaginous green leafy vegetable, sometimes called Jew's mallow.

Since coriander seeds tend to brighten the flavor tone of a dish, they are a good companion to **cumin**, which has a darker, dustier taste, and the two are often used together. The combination is popular in Mexican and Tex-Mex cuisine—it's essential in **chili powder**. Middle Eastern cookery also favors this pairing. Elsewhere, the two spices are a feature of Indian **chaat masala** and the Parsi **dhansak masala**, and Indian shops stock a ready mix of the two spices, in whole and powdered versions, called *dhana-jeera*.

coriander leaf/cilantro/Chinese parsley: The leaves of the coriander plant resemble flat-leaf **parsley**, but coriander's color is a lighter green and the leaves are more deeply indented. Most important, coriander leaves have a more potent aroma and a totally different taste than parsley. Descriptions of the taste of the leaves vary, since people either adore or abhor it. For its many fans, the unique taste is cool and refreshing; while the opposing group considers it unpleasantly brash or even "buggy." (The claim is made that the name "coriander" derives from the Greek word for bedbug; but those who have encountered both herb and insect report that there is no similarity of odors.)

The pretty, lacy leaves are often used as garnishes or mixed with other fresh ingredients in salsas or chutneys. The popular South American condiment *chimichurri* and the Moroccan *charmoula* are both sharp acidic sauces based on coriander leaf and **garlic**, a winning combination of flavors, as is coriander and onion. Coriander leaves have a special affinity for fresh green **chiles**, and the two together create some exciting, brilliant green dishes.

Coriander leaves are not much used in Europe, but they are rapidly gaining in popularity in North America, mostly through the influence of Southwestern cuisine. Hence, in the United States they are most often known by their Spanish name, cilantro.

Latin America adores this leafy herb, as does the Middle East. India and southern China consider it a staple. While Indonesians tend to concentrate on coriander seeds, Filipinos love the leaves, and the Thais eat seeds, leaves, stems, *and* the fragrant roots of the plant as well, usually grinding the latter for curry pastes and other dishes.

Coriander leaves do not dry well, as much of the flavor seems to evaporate along with the moisture. Similarly, cooking robs the herb of some of its essential oils. For that reason, the leaves are generally added to hot dishes at the end of the cooking time. However, some Indian dishes cook the leaves for one aspect of their flavor and then add a little fresh just before serving for another aspect. You might like to use the stems to cook with, picking them out before serving, and then adding the leafy tops. The stems can be held in the freezer until needed.

If you wish to tone down the strong taste of cilantro, mix in some flat-leaf **parsley**. If you cannot get cilantro, or you do not like it, substitute a little **oregano**.

Fecund Mother Nature proffers a number of unrelated herbs with an intense aroma resembling that of cilantro, for which they are prized as a seasoning and used in the same way as cilantro in various parts of the world. Two important examples are discussed below.

culentro/cilantro ancho/saw-leaf herb/pak chee farang (*Eryngium feotidum*): This New World native has a definite cilantro aroma, stronger than that of coriander itself. But the two plants look entirely different. Culentro leaves are broad and flat, approximately three inches long and one inch wide, with almost prickly saw-toothed edges. They are used as a seasoning throughout much of Latin America, but especially in the Caribbean where they are particularly popular in rice dishes.

Asian cooks also use this herb. When they call it *pak chee farang*, they are saying "foreign coriander." In Vietnam it is named *ngo gai*. Lots of cilantro is the obvious substitute for this herb.

Older, larger leaves of culentro can become quite tough. For these, cut out the hard central veins and slice the leaves into narrow strips.

rau ram/laksa leaf/daun kesom/Vietnamese coriander/polygonum (*Polygonum odoratum* and other *Polygonum* species): A distinctive dark arch usually—but not always—marks the center of each lance-shaped, green leaf. The leaves have an aggressive smell and a complex taste that finishes with a surprising sensation of heat on the tongue. The fact that this herb is also sometimes called Vietnamese mint indicates just how unusual the flavor is.

This intensely fragrant herb is never cooked. It is served raw in Southeast Asian salads and as a garnish for soups and other dishes. This herb is sometimes called laksa leaf because it traditionally tops the rice-noodle soup, *laksa,* a Nonya specialty of Singapore. ("Nonyas" are the Straits Chinese women, whose culture and cuisine blend Chinese and Malay influences.)

cumin (*Cuminum cyminum*): Because of their small size, the dried fruits of this plant are commonly called "seeds." Each cumin seed—technically half a fruit—is tan with lighter-colored ribs run-

ning lengthwise. In appearance but not in flavor, cumin seed resembles its botanical cousins in the parsley family: **anise**, **caraway**
and **fennel** seeds. Cumin has a unique sharp, earthy flavor with sour
overtones. Its taste is dominant and persistent, and the spice should
be carefully measured to control the final effect.

As a plant best suited to warmer climates, cumin is associated with
the cuisines of warmer countries, especially India, North Africa, and
the Middle East. Cumin is basic to Indian cuisine, marrying well with
vegetables and pulses. The Anglo-Indian soup mulligatawny includes
cumin. The seeds flavor many a **curry powder**, and are an essential
ingredient in the various Southeast Asian curry pastes. Cumin is the
darling of Moroccan cooks. It is the perfect companion for lamb, and
sometimes a morsel of hot roast lamb is simply dipped into a little
mound of ground cumin, then into salt, on its way to the mouth. In the
Middle East, cumin may be used in every type of dish except sweets,
but it is especially favored for fish, meats, and lentils. It is frequently
paired with **coriander** seeds. The popular Persian spice blend **advieh**
usually includes cumin. Cumin seeds also tickle tastebuds in Holland,
where they are added to some cheeses.

Assertive enough to stand up to **chiles**, cumin has won a place in
Mexican cuisine, where it is used, with discretion, to season the stronger-flavored meats and the tripe dish *menudo*. This spice often teams
up with dried **oregano** and a pinch of **cloves**. And cumin contributes to
the distinctive taste of **chili powder**.

A sprinkling of ground cumin wakes up any egg dish, from poached
to deviled. Cumin butter, a blend of toasted, ground seeds and soft
butter, makes a delectable accompaniment to fish. When plain vegetables or plain rice threaten to be boring, sprinkle a few teaspoons of
cumin seeds over them. Add a pinch to dressings for green salads.

There is a darker, smaller variety of cumin seed called black cumin,
which is sometimes used in Central Asia, Russia, and India. Its sweeter, mellower flavor is often preferred for pilaf, *manti*, and other dishes.
Ordinary "white" cumin can always be substituted for black, although
the flavor of the white variety is not as subtle. But be careful: the name
"black cumin" is often misused to mean **nigella** seed.

Ground cumin loses its flavor rapidly; you will get better results if
you buy whole seeds, and toast and grind them as needed.

curry leaf/kari patta/karuvepillai (*Murraya koenigii*): Curry powder is *not* made from powdered curry leaves, but the leaves of this Indian shrub or small tree are a favorite ingredient in curries from the south of India and along the west coast of the subcontinent; they are not used so much in the cooking of northern India.

You can probably find curry leaves, dried or fresh, in an Indian grocery shop. Fresh leaves are better and if you find them, buy a few extra to put aside in the freezer (see Freezing Herbs in the "Culinary Practice" chapter). When buying dried leaves, check them first to see if they have any aroma left—they should have a very delicate clean smell.

You can use curry leaves like **bay leaves**, but in quantity: depending on their freshness, a couple of sprigs with a total of six to ten leaflets are needed for a dish serving four. Remove the sprigs before serving.

Curry leaves, with stems removed, can also be sautéed lightly in oil, over medium-high heat, along with the other spices called for in a recipe. This method should be chosen when making a dry curry—that is, one without much "gravy" in which the leaves could simmer—and also for the savory spiced-rice dishes of south India. In these dishes, the leaves are not picked out before serving; rather each diner can either push them aside or munch them up, according to preference.

Note that curry leaves are sometimes called "sweet neem" (also spelled *nim*), but don't confuse them with (bitter) neem, *Azadirachta indica*, an important tree of India whose twigs make efficacious natural toothbrushes and whose slender leaves—with a longer, more pointed tip than curry leaves—provide protection from insects for rice, textiles, and organic crops.

D **dill** (*Anethum graveolens*): Dill is one of those generous plants that supply us with both seed and leaf as seasonings, but when a recipe calls for just plain "dill," you can be quite sure it means the leaves. The feathery leaves are also called "dillweed"—a libelous label for a refined flavor.

Use the fresh leaves uncooked, adding them to hot dishes as the final step in preparation. Fresh leaves are far superior to dried, which retain very little flavor. If you must use the dried leaves, allow them some time to soak in a liquid, either in the process of cooking the dish or, in the case of soups and sauces, after the dish is prepared. Use about half as much dried dill as you would fresh, and then taste to see if you want to add more.

Freezing is a better way to preserve dill than drying; however, though frozen dill will be flavorful, its texture will be mushy, making it unsuitable for use as a garnish or salad ingredient.

Dill has a penetrating fresh, clean flavor that can overwhelm a dish; often a wispy plume of leaves laid on the plate as a garnish is sufficient. At other times, when pairing the herb with another strong taste, you will want a larger quantity of dill.

Dill enjoys huge popularity in a wide arc curving from Scandinavia through the former Soviet Union and into Persia and Turkey. But step out of this region and you find the herb reduced to obscurity, known almost exclusively from dill pickles.

Dill is the most popular herb in Scandinavian cuisine. The leaves are excellent with all kinds of fish, and are an essential part of the process of preparing *gravlax*: fresh salmon filets and great quantities of dill are pressed under weights for a few days, along with coarse **salt**, sugar, and white **pepper**. Sophisticated gravlax recipes sometimes call for a sprinkling of cognac, as well. This delicacy is traditionally served with a mustard-dill sauce.

Having a natural affinity for fish, pickles, sour cream, and cucumber, dill dresses and garnishes numerous dishes on the smorgasbord. A favorite meat dish in Sweden and Finland is braised cubes of lamb accompanied by a white sauce made with lamb stock and chopped dill; the sauce can be made sweet and sour, if you like, with sugar and vinegar.

Late-summer nights in these countries are the occasion for dinner parties featuring dill-flavored crayfish. The crayfish are prepared hours ahead by boiling them briefly, each quart of water in the pot being spiked with a tablespoon of dill seed and a small bunch of fresh dill leaves; the pot is then covered and the crayfish allowed to cool slowly in the dill water until time for the party. These bright red shellfish are spectacular when garnished with large, yellow dill-flower clusters. Dill flowers are also edible, with a mild flavor similar to that of the leaves.

A treasured dish in southern Russia and Ukraine is *borsch*, a rich soup based on beef stock with beets and cabbage, and a choice of other vegetables such as potatoes, carrots, onions, and tomatoes; there is an array of seasonings at the discretion of the cook, but many cooks in that region are of the opinion that without dill, it isn't borsch, it's just soup! Serve borsch either hot or cold, with a spoonful of sour cream in each bowl, and lavished with bright green chopped dill.

Persian cuisine also favors dill. Cooks in Iran make a standard vegetable rice with lima beans (or fava beans) and generous quantities of finely chopped dill. They also include dill along with **mint**, **tarragon**, scallions, rice, and split peas in making stuffed grape leaves. *Sabzi polow*, an herbed rice, combines boiled rice with an equal amount of herbs: **chives** or leeks, **coriander** leaf, **fenugreek** tops, and dill. Note, however, that dill does not appear on the traditional platter of fresh herbs that begins a Persian meal.

Fresh dill is also used in numerous ways in Turkish cuisine. It is an indispensable garnish for all kinds of vegetable dishes. Chopped dill is mixed into the filling for cheese *börek*—fried pastries filled with soft white sheep's cheese—and it flavors *cacık*, a refreshing cucumber and yogurt mixture. Fresh dill accompanies a cold vegetable salad of thick slices of boiled **celery** root (celeriac) dressed with **lemon** juice and olive oil; and dill is a stunning final touch on the famous Turkish rice pilaf enhanced with onions, currants, pine nuts, and (optionally) bits of lamb's liver.

In America, the most famous use of dill is in good and cold, crunchy, sour cucumber pickles. Sprigs of fresh dill are packed in the jar with the cucumbers, and of course dill seed is included among the **pickling spices**. This combination of sharp vinegar, cool cucumber, and warm dill is so successful that it has even been used as a novelty flavor for potato chips.

American cooks are learning to take advantage of dill's lively flavor in other foods besides pickles. Just a sprinkling of chopped fresh dill turns the most everyday peas-and-carrots dish into a specialty for guests. It also lifts a coleslaw out of the ordinary; use it as a major ingredient, adding as much as a cup of chopped dillweed to a medium head of cabbage in the slaw. A light sprinkling, finely chopped, atop a bowl of tomato soup has a dramatic effect in both color and flavor. A quick and easy cold dill sauce for seafood, chicken, or raw vegetables can be whipped up by adding chopped dill to mayonnaise along with a little lemon juice and chopped **parsley** or other herbs, as desired.

dill seed: The seeds—actually fruits—of dill also constitute a useful spice. These distinctive flat ovals are light brown with a narrow white or beige edge. The taste of dill seed is stronger than that of dillweed and, unlike the leaves, it resembles the taste of **caraway seed**, to the point that the two seeds can be substituted for each other in equal quantities.

Almost all the dill seed available commercially in the United States is grown in India. Indian dill, *Anethum sowa*, is a different species than the one grown in the United States for its leaves or for dillweed oil. (If you buy dill seed in an Indian grocery, it may be labeled *suva*.) The flavor of Indian dill seeds has fewer of the caraway flavor notes, but they can nonetheless be used in exactly the same way as seeds of *Anethum graveolens*.

Dill seeds make an excellent flavored vinegar (see the "Flavor Combinations" chapter) and are a good addition to **crab boil** spices. Their flavor is very good with boiled carrots, potatoes, and other root vegetables such as turnips, rutabagas, and beets; simply add a tablespoon of seeds to the cooking liquid.

For a more dramatic effect, dill seeds can be added to cooked vegetables just before serving; for example, the seeds are attractive and delicious scattered over a potato salad. But be careful when using dill seeds in this way: the raw seeds are unpleasantly tough to chew, and even when you grind them in a mill or a mortar, the remaining bits are still hard. You can solve this problem by boiling up a small quantity of whole seeds in enough water to cover, and letting them simmer for half an hour. Strain them out and dry them on a kitchen towel. (The remaining liquid, by the way, is the famous "dill water" often prescribed for colicky babies and others with upset digestions.) The cooked seeds are tender enough to be used whole, or you may grind them into smaller pieces if you prefer.

E epazote/Mexican tea/goosefoot/American wormseed (*Chenopodium ambrosioides*): In addition to all the foregoing names saddling this native American herb, it is occasionally referred to as "pigweed" and "Jerusalem oak," and in Spain it is *pazote*, a name that also applies to the herbal tea made from its leaves. The narrow, pointed, serrated green leaves waft a potent odor, and their flavor is sharp, minty, and somewhat mothballish. There is also a purple variety of this plant, which some people prefer for its milder flavor; others disdain it for the same reason.

Epazote is popular in the cooking of central Mexico and on southward through Central America. It is considered essential in black beans, both for enriching the flavor and because it is believed to take the wind out of beans. Beans of all kinds, mushrooms, tortilla dishes such as *chilaquiles* (mixtures of tomatoes, chiles, and strips of tortillas with beans, chicken, or cheese) and *quesadillas* (cheese, sauce, and other ingredients in tortilla "sandwiches") all benefit from a light application of epazote. Its assertive taste is good with spicy sauces of all kinds; add it sparingly to salsa. This herb is also superb with fish and shellfish. Try a few finely chopped leaves in a western omelette.

For beans and other cooked dishes, add a sprig of fresh epazote to the pot for the last quarter hour of cooking. Remove the sprig before serving. If you have only the dried leaves, use about one teaspoon to replace a sprig of fresh leaves. For quesadillas and uncooked sauces, add a sprinkling of finely chopped fresh leaves.

While fresh epazote is clearly superior, it is not always available. The leaves do dry sufficiently well for cooking or for brewing into tea. If you cannot get epazote, substitute a good strong **oregano** or fresh **coriander** leaves (cilantro).

F fennel (*Foeniculum vulgare*): Both the seeds and the feathery leaves of fennel are used as seasoning, and the bulbous base of one variety is cooked as a vegetable.
Popularly called seeds because of their small size, fennel seeds are actually the dried, ripe fruits of the plant; to be precise, each one is

half a fruit. They are greenish-brown in color, slightly curved and deeply ridged. Superficially, they resemble **anise** seeds, but fennel seeds are larger and more elongated, and are not as likely to retain a wisp of stem. However, the aromas and flavors of these two spices are similar to some degree, depending on the variety of fennel. Some fennel seeds taste strongly of anise, while others have only a mild hint of that flavor. The taste of fennel seeds also has a distinct bitterness not found in anise seed.

Fennel seed is a favorite spice in India and China, as well as the countries in between. It is important in Chinese **five-spice powder**. In India, the seeds are given a brightly colored candy coating and served after meals as a digestive.

Confusingly, Southeast Asian cooks sometimes refer to fennel seed as "sweet cumin." Fennel seed and **cumin** seed are similar in appearance—although fennel is generally larger—but they are very different in flavor. Whereas fennel is slightly anisey with a bitter note, cumin is penetratingly earthy with a sour overtone.

In Europe, fennel seeds are a popular topping for breads and pastries, and are also incorporated into many Italian sausages. Sometimes fennel seeds appear in **herbes de Provence** in place of **tarragon**. They are good with carrots and winter squash.

Fennel seed has been called "the fish spice," being the perfect spice especially for the oily fishes. Shrimp and crab also profit from fennel seeds, and they are a good choice for a **crab boil** blend.

How to use fennel seeds with fish? Begin by toasting the seeds lightly. About one-fourth to one-third teaspoon per serving is ample. If the fish is poached, toss the whole seeds into the liquid. If the fish is broiled or grilled, crush the seeds coarsely and make a compound butter (see the "Flavor Combinations" chapter) to slather over the hot fish as soon as it is cooked. If the fish is battered and fried, crush the seeds finely and incorporate them into the batter. And for baked fish, sprinkle the seeds under the fillets in an oiled or buttered baking pan; add a small splash of wine or milk to keep everything moist and tender. Serve with the baked seeds and juices poured over the fish.

All parts of the fennel plant taste of anise, but the leaves least of all. The feathery light green fennel leaves can be used like dillweed,

but with a different flavor effect: They are much milder and have that slight anise taste which is entirely missing in the dill. They make a beautiful, fresh-flavored topping for fish dishes or an attractive and tasty bed on which to serve fish. They can be used with any vegetable dish that needs a little brightening; try them with sauerkraut.

The azoricum variety of fennel—sometimes sold as *finocchio*, Florentine fennel, anise fennel, or just plain anise—is cultivated for its large "bulb," which is really the bulge formed by the overlapping of its expanded leaf stalks. This vegetable is usually braised, but it can be eaten raw in salads or served in slices spread with soft cheese, like celery stalks. Chopped fennel leaves are an appropriate garnish for dishes that contain the fennel bulb.

fenugreek (*Trigonella foenum-graecum*): The odd species epithet of this plant translates as "Greek hay," suggesting that it was long ago grown as fodder. This legume certainly provides a great deal of nourishment for livestock, although some people feel it introduces an off taste into the milk of animals that eat it.

fenugreek seeds: The plant bears a slender pod, tapering to a point, which contains ten to twenty fawn-colored, flat, rectangular seeds, each with a diagonal groove on one side. These seeds have a forceful, bitter taste with nuances that might be described as bitter maple. In fact, fenugreek seeds have been used to make artificial maple flavoring. Their powerful, lingering smell may be recognized as the predominant scent in many prepared **curry powders**, and it can override all other aromas in an Indian spice shop. The spice is tamed a bit by cooking, however, and it blends in nicely with the other seasonings in a finished dish.

Because the seeds are quite hard, it is advisable to buy commercially ground fenugreek when you want a powder. If you are grinding the whole seeds yourself, toast them lightly first; this will bring out their flavor and simultaneously make them somewhat easier to grind. It is all too easy to scorch the seeds when you toast them; a barely perceptible change of color is sufficient. Overheating the seeds can spoil their flavor.

Fenugreek is a particular favorite spice in south India and Sri Lanka, where it is not limited to curries but may also be found in chutneys, lentil dishes, pickles, and vegetables; the flavor is especially felicitous with potato, eggplant, and cauliflower. Fenugreek is important in **sambar powder** and in the **panch phoran** seasoning mix of Bengal.

Pastirma (or basderma), the spiced dried beef so loved in Armenia and Turkey, is cured with fenugreek powder and **garlic**, enriched with **allspice**, **cumin**, black **pepper**, **cinnamon**, **nutmeg**, and **cloves**, and colored a rich red with **paprika** and ground red **chiles**. This spice mixture is called *chaiman* (or *chemen*), as is the fenugreek itself. The chaiman mixture, moistened with water, is sometimes spread on bread for an unforgettable snack.

An extraordinary condiment is made with fenugreek in Yemen and in the Jewish community of Calcutta. *Hilbeh* (again the condiment is called by the local name for fenugreek) is made by pouring ample boiling water over about two tablespoons of ground fenugreek in a mixing bowl, and letting it sit undisturbed for several hours or overnight. When the excess water is carefully poured off, the fenugreek will have gelled into a thick mass. Beat this jelly at high speeds, while adding about half a cup of water drop by drop, and it will rise up into a permanently foamy sauce. Then stir in your choice of spices; the basic seasonings for hilbeh are a pinch of **salt**, a little **lemon** juice, and some cayenne pepper, but it can be made as elaborate as you like with the addition of **garlic**, **ginger**, **caraway** or **cardamom** seeds, chopped tomatoes, finely chopped **chiles**, and **coriander** leaves. Refrigerated in a sealed container; hilbeh will keep for a week.

Hilbeh is served like a chutney or a jelly with meats or vegetables. A dollop may be added to a bowl of chicken or other soup, or it may accompany one of the Jewish Sabbath stews.

If you like the flavor of fenugreek, try adding a tablespoon of the seeds to thicken a stew—they will swell and become mucilaginous when cooked—or use ground fenugreek to flavor a marinade for shish kebab cubes and other meats to be grilled.

fenugreek leaves/methi: In India, the fragrant green leaves of this plant are eaten with great enthusiasm. Fresh methi, cooked as a green, is a favorite served with fish. Bunches of the fresh leaves are also chopped and put into curries and sambars (see sambar powder in the "Flavor Combinations" chapter). Highly seasoned with **turmeric**, **cumin**, **ajowan**, fresh **ginger**, and **chiles**, the leaves are mixed with cornmeal and fried as fritters. If the leaves are only briefly cooked, the stems should be removed. Fresh methi can be found in season at Indian shops; dried leaves are available all year round. If possible, choose young, fresh fenugreek leaves for your cooking; they have a much better flavor than the older ones. One way of ensuring a supply of good fresh fenugreek is to grow it yourself: almost any store-bought seed will grow when it is planted.

Crushed dried leaves, with larger stems removed, are sprinkled over curries or cooked vegetables just before serving for an evocative flavor. This garnishing is particularly effective with buttered carrots and potatoes. Finely crushed, the leaves may be mixed into the dough for nan and other Indian breads.

Fenugreek leaves are also important in the cooking of Iran. Fresh or dried, they are used to add strong flavor to stews, soups, and *ashes*—thick, nourishing, main-course soups. For these dishes, the leaves are usually first sautéed in oil, along with other green vegetables and herbs such as **parsley**, **celery**, scallions, leeks, or spinach, before they are added to the dish; this allows them to be added midway through the stewing process, so as not to be overcooked. Fenugreek is featured in the famous herb stew *ghormeh sabzi*, a rich, dark green mix of herbs, beans, dried **limes**, and some meat.

The dried leaves of another species of fenugreek, sweet trefoil, *T. caerulea*, are popular in the cuisine of the Republic of Georgia. These leaves are an important part of the traditional Georgian spice blend **khmeli-suneli**, and an occasional pinch of the powdered leaves is added to soups and stews.

fenugreek sprouts: If you enjoy fresh sprouts, try sprouting fenugreek seeds. Soak one or two tablespoons of seeds in warm water for a few hours, then lay them on a soaking-wet paper towel in a sealed, opaque container made of glass or plastic, *not* metal, wood, or unglazed earthenware. Put them away in a warm part of the room, out of drafts. Each morning and evening (and at midday, if possible), change the towel for a fresh wet one. Within four days, the sprouts should be ready. Don't let the sprouts grow very long: one-fourth inch is ample, and sprouts more than half an inch long will be too strong-flavored to eat raw. A few small sprouts add an exciting crunch to a vegetable or bean salad, or to a stir-fry. Fenugreek sprouts may be added to mustard or mayonnaise for a roast beef sandwich, or folded into hot cooked rice to make a memorable pilaf.

G galangal/greater galangal/galanga/Siamese ginger (*Alpinia galanga*): Some botanists prefer to classify this plant as a member of the genus *Languas*. This appetizingly aromatic rhizome is cultivated throughout Southeast Asia, where it is extremely popular. The spice is called *kha* in Thailand, *laos* in Indonesia, *languas* in Malaysia, and *gieng bot* in Vietnam. (These words appear in a variety of spellings; for example, *kha* may be written *ka, khaa,* or *kaah; languas* may be *lengkuas;* and *gieng* is sometimes *rieng*.)

The large, fleshy rhizome bears a superficial resemblance to **ginger** root but, unlike ginger, it has prominent thick stems with a pinkish cast at their base, and its pale skin is made up of bands with darker edges, creating a sort of zebra pattern. Galangal's perfume is less penetrating than ginger's, but its taste is more peppery.

Today galangal is grown in Hawaii and shipped fresh to Asian shops in the continental United States. It is also widely available as a powder.

Galangal lends its distinctive taste to Southeast Asian curries, fish, and other savory dishes. It is important in the Indonesian condiment *sambal bajak*. The classic Thai chicken soup, *tom kha gai*, features galangal, along with almost every other flavor of the Thai palette—**coconut** milk, **lemon grass**, kaffir **lime** leaves, **chiles**, fish sauce, and **coriander** leaf.

Galangal has also established itself in certain enclaves in the Middle East, where it is called *khulinjan*. In Saudi Arabia, its warm, earthy flavor is traditionally incorporated into meaty soups; and in Morocco, galangal is one of the myriad ingredients that make up the spice mixture **ras el hanout**.

Galangal deserves a wider audience. Once you try adding a little to consommé, you'll never want to be without it!

rhizome/krachai/lesser ginger (*Kaempferia galanga*): This is another saporous rhizome used in the cuisines of Southeast Asia. This plant has been the subject of much confusion in the West simply for lack of an established name. *Krachai* is the Thai name for the spice; it is known as *kencur* in Indonesia. (Again, the Thai has been variously transliterated into English: you may see *grachai* or *kachai*. The Indonesian name is sometimes written with the spelling *kentjoer.*) As to its English name, frustrated exporters often label the spice simply "rhizome." The plant has been called false galangal in India. Occasionally, it is lesser ginger. Some food writers apply the name "lesser galangal" to this rhizome, but historically that term has denoted the small (hence the name) brown root of the medicinal plant *Alpinia officinarum*, a native of Hainan Island and the nearby Chinese mainland.

Krachai grows in "hands" of long, slender fingers covered with a thick brown skin. Under the skin is an orange-yellow core of juicy flesh. Be aware as you work with it that the juice can impart a yellow stain to clothing.

This spice is especially popular in Thailand; it is much stronger in flavor than greater galangal, and Thai cooks are particularly fond of its sharp taste. (One Indian cook plausibly compared its taste to that of **amchur**.) It is also an indispensable ingredient for the popular crisp vegetable salad *karedok*, a specialty of Java.

Whole rhizomes of krachai are imported from Thailand, vacuum-packed and frozen. You can also find packets of dried slices and little bottles of powder.

Both of these two spicy rhizomes, krachai and galangal, are used in the same way in the kitchen, except that you cannot have so free a hand with the more pungent krachai. Peel the fresh rhizomes and chop them fine. The pieces may be further pounded with other spices to make Thai curry pastes.

The dry powders are inferior in flavor to the fresh or dried rhizomes, but they are still entirely usable and will provide the unique tastes needed for Southeast Asian cooking. One teaspoon of powder is approximately equivalent to a quarter-inch-thick slice of the fresh rhizome, or a dried slice one eighth-inch thick.

Tempting though it is to substitute the similar-looking ginger for galangal or krachai, their tastes are really quite different. You might try substituting half as much ginger plus a generous amount of freshly ground black **pepper** for the rhizome; temper the combination, if you wish, with a pinch of ground **cardamom** seed. Or you may decide that it is better to omit these spices altogether. Without galangal or krachai, your Southeast Asian dishes will still be good, although to native eaters of these cuisines they may seem to have a foreign accent.

The English word "galingale" is applied to both greater galangal and lesser galangal, both of which were commonly imported into Europe during the Middle Ages. Galingale was a popular flavoring in sauce for fish, and was also used with meats. The name was also given to the domestic substitute for imported galangal: the fragrant roots of a local sedge, *Cyperus longus.*

garlic (*Allium sativum*): This robust member of the lily family grows a bulb, compounded of several separate "cloves," each wrapped in a papery skin that is usually white but may be pinkish in some varieties. The size of the bulb varies also. Elephant garlic, for example, is a mild-flavored variety with huge cloves. Treat it just like ordinary garlic.

The taste for garlic varies enormously. Today garlic is generally socially approved as a whole-hearted, free-spirited, anti-elitist sensual pleasure, in contrast to the prevailing attitude in America a few decades ago, which regarded garlic breath as a serious faux pas if not a socially debilitating disgrace. Those who liked it permitted themselves only a suspicion of garlic. It was recommended to rub a cut clove over the inside of the salad bowl to obtain just the aroma of garlic in a salad. Taken to extremes, the rule was simply to "rub the garlic on the cook."

However, it is clear that some people are more affected by garlic than others. Very fair-skinned people, especially, sometimes suffer what have been described as "garlic hangovers" (headache, dehydration, slight nausea), and anyone who cooks for such people needs to respect their sensitivity. One way to produce only a gentle hint of garlic flavor is to drop the unpeeled cloves into boiling water for two or three minutes before using them in the recipe. You can also choose elephant garlic, which is milder, or use garlic **chives** instead. **Asafetida** is an excellent substitute for garlic; use one-fourth teaspoon of powdered asafetida in place of one small clove of fresh garlic, crushed, or according to taste.

When selecting garlic bulbs or heads, do not buy any that have already sprouted. (The same applies to onions and shallots.) The bulbs should be firm to the touch, not hollow, soft, or loose, and have a dry papery covering. Heads of garlic will last for weeks if they are kept cool and dry, but do not keep garlic in the refrigerator. A ceramic garlic keeper, with several large air holes, really does work. Individual cloves can be removed from the head when you are ready to use them, but don't separate the head beforehand because the cloves dry out much faster.

It is *not* a good idea to store garlic cloves in a jar of oil; this is a dangerous practice because it creates the perfect anaerobic, low-acid environment enjoyed by the bacteria that cause botulism. Even if the jar of oil is kept in the refrigerator, some types of botulism bacteria may be able to survive and produce their deadly toxins. (For more on botulism, see Flavored Oils in the "Flavor Combinations" chapter.) Commercially prepared garlic oils are safe because they are made with equipment capable of sterilizing the product. Similarly, commercially sterilized chopped garlic in a jar, mixed with oil and preservatives, is safe from botulism, although it has a definite "canned" taste.

To separate the cloves of a head of fresh garlic, lay it on a table and smack it sharply with the side of your fist. To peel the skin off a single clove, cut off the stem end and lay the clove on a board with the side of a large knife resting on top of it; hit the side of the knife with one quick blow and the papery peel can then be easily lifted off the clove. You can, of course, also use any of the myriad gadgets designed for this specific purpose; they range from simple little wooden implements shaped like mushrooms, with which you are meant to press on a clove until it pops its skin, to electronic devices that turn out one peeled clove after another in rapid succession.

The peeled clove should be creamy-white and unblemished. Cut off any brown spots. If your garlic is not very fresh, you may find that it has already begun to sprout inside the clove. This sprout has an unpleasant, bitter taste, and you will enjoy your garlic much more if you "degerm" it by cutting into the peeled clove and removing the pale green sprout residing in the middle.

A clove may be used whole, cut across into attractive thin slices, or laid under the blade of a chef's knife and bashed. The bashed clove may be further scraped across the cutting board until it is a fine juicy mush. A fresh, fat peeled clove can also be cut very precisely into tiny cubes with a paring knife by first making horizontal slices, then cutting vertically in two planes, much as you might dice an onion. These attractive, regular cubes are ideal for that exquisite garnish and topping, **gremolata**.

The intensity of the flavor of fresh garlic depends upon how you handle it. A general rule of thumb is, the more you do to garlic, the more it does to you. A whole garlic clove will season a dish far less than the same clove cut into pieces, and a crushed clove will be still more intense. But no matter what your taste in garlic, be sure to avoid that ultimate mauler, the garlic press. This contraption manages to squeeze out for you all the least attractive flavor compounds in the bulb and to keep to itself the sweetest ones. (But don't throw your garlic press away: it is the ideal implement to use with fresh **ginger** root. Perhaps we should rename it a "ginger press.")

The intensity of garlic in a dish also increases with time: a dish with just the right amount of chopped garlic tonight will be almost radioactive with it when served as a leftover tomorrow. If you plan to make a garlicky dish a day ahead, it is better to use a whole clove or two and then fish them out when they have done their job to your satisfaction; mark them by sticking them with a toothpick or small skewer so you can easily find them again.

Be careful with the many recipes that begin with sautéing minced garlic in oil or butter: the heat should not be high and the garlic should only reach a pale golden color. It will have a dreadful acrid flavor if it is browned too much. Never sauté onions and garlic together—as so many recipes instruct you to do—because the garlic will be overcooked long before the onion is ready. Sweat, sauté, brown, or caramelize the onions first, as your recipe requires, and then add the garlic in the last few minutes of the process.

roasted garlic: Roasted garlic is surprisingly mild. Remove the loose parchment around a head, but do not peel it. Slice off the tips, just exposing the flesh in each clove. Set the head upright in a dish and drizzle a little olive oil over it. Bake at 350°F for fifty minutes or longer, depending on the size of the head, and baste it occasionally with the oil. Allow to cool enough to handle and serve whole. Try serving roasted garlic with a garland of **rosemary** sprigs; the herb's aroma alone will flavor the experience of eating the bulb. The individual cloves are pulled off the roasted bulb and the roasty-smoky, nearly liquid garlic is squeezed out onto a slice of good bread or toast.

It is not necessary to roast an entire head of garlic. A few cloves can be dropped, unpeeled, into the pan around a roast or a hen, or even in a separate small dish; in both cases they should be basted. Having a few extra roasted garlic cloves available in the refrigerator is convenient when you are making a dip or sauce; just squeeze out the smooth, flavorful paste from the roasted clove, mash it with a fork and mix it in thoroughly.

pickled garlic: Garlic is pickled in Thailand and eaten as a snack; it is also used for cooking, especially with noodles. You can easily pickle your own peeled garlic cloves by boiling them for five minutes completely covered with one and one-half cups water, one-third cup distilled **vinegar**, one tablespoon sugar and one teaspoon salt. Be sure to use a nonaluminum pan. After boiling, set the garlic cloves in a jar and pour the hot liquid over them to cover. Refrigerate for two weeks.

Forget dehydrated garlic powder! It simply does not do justice to this luscious seasoning. If you insist upon using garlic powder, measure out one-fourth to one-half teaspoon for each fresh garlic clove that you might be using instead. And store your garlic powder carefully—in a tightly closed container kept in a cool, dry, dark place, for no longer than six months—because it can turn rancid.

garlic salt: Juicy, fresh, homemade garlic salt is a snap. Peel one clove of garlic and mash it in a mortar with four to six teaspoonfuls of a good **salt**; the exact amount will depend on the size of your clove and on how garlicky you want your dish to be. Store any leftover in the refrigerator, and don't keep it for more than a couple of weeks.

ginger (*Zingiber officinalis*): This spice is often called "ginger root," but technically what we use is the rhizome, or underground stem, of the plant. The knobby pieces of rhizome are also referred to as "hands" or "races" (derived from the Latin word for "root") and it has been suggested that it is from races of ginger that we get our somewhat dated word "racy," meaning something which is a bit risqué, or piquant and spicy. Occasionally, a piece broken off from a hand of ginger is referred to as a "finger."

Ginger is indispensable in all Asian cookery, and is a favorite spice in the West as well. Gingery baked goods of all kinds have been so popular that they are sometimes called for as an ingredient in other recipes; for example, crushed gingersnaps are frequently used as a thickener in the gravy for sauerbraten, and the dry crumbs of stale gingerbread make an excellent breading for meats. An easy and interesting pie crust can be made by mixing gingery crumbs from g-snaps or g-bread with melted butter and pressing them into a pie plate.

Try adding a little ground or grated fresh ginger to melted butter for basting poultry or roast meats.

fresh ginger/green ginger: In Chinese cuisine, the way that fresh ginger is prepared is immensely important; the cut of the root—grated, sliced, slivered, or shredded—alters the texture and therefore the character of a dish. For most Western recipes, however, the important thing is to have the desired strength and pungency of flavor and to avoid any unpleasantly tough ginger fiber.

Purchase fresh ginger in small amounts. Leftover pieces can be kept, well-wrapped, in the refrigerator for as long as two weeks. Freezing preserves the flavor but makes the root mushy. When buying fresh ginger, look for firm and weighty pieces; never buy shriveled roots. Smooth skin is a good sign, though some varieties have rougher skin than others. Generally, pungency increases with age, and so does fibrousness.

You can usually peel fresh ginger root with a vegetable peeler. If the ginger is young and tender, the peel can simply be scraped off with a knife. If you have a particularly recalcitrant piece, soak it in water for about an hour; it should then peel easily.

GARLIC

Since ginger root of many different varieties is imported into the United States from a number of different countries and at all times of the year, it is a good idea to slice off a small piece of the root to taste it and determine its fibrousness before dealing with it in one of the following ways:

Slice. For soups or sauces, peel the root and slice it into disks about one-fourth inch to one-half inch thick, cutting at right angles to the fiber. Usually three or four of these disks will suffice to flavor two cups of liquid, but the amount needed depends on the strength of the ginger, the type of dish being prepared, and your own fondness for the taste. This is a good way to use a tough, fibrous piece of ginger, because the disks will be removed before the dish is served.

A slice of ginger is commonly fried in the hot oil in a wok before the stir-fry ingredients are added. Any member of the cabbage family, such as brussels sprouts, broccoli, broccoli rabe, Chinese broccoli, cabbage, or kale, is improved by cooking with a few slices of fresh ginger. Serve the vegetable seasoned with soy sauce rather than salt, and brighten the flavor with a squeeze of lime juice.

Sliver (or julienne). Peel the root and slice it about one-eighth inch thick; then stack up several slices and cut through the stack in parallel cuts, about one-eighth inch apart. These slivers are decorative as well as flavorful and are meant to be served with the dish, although they are not necessarily to be eaten.

Shred. Choose tender young ginger for shreds. Slice a piece of peeled ginger root as above into whisper-thin slivers, as fine as you can manage. A cleaver works very well for this; otherwise, use a large chef's knife. Serve ginger shreds as a topping and devour.

Chop (mince). Cut ginger root into slivers as above, then turn them 90 degrees and cut a second time, more or less dicing the root. The fineness of the dice depends on the nature of your recipe. A tender chopped ginger can be left in the dish and eaten. Chopping is also a good way to treat tough, fibrous old ginger for marinades and syrups that will be strained later.

Grate. Japanese porcelain ginger graters with raised spikes work better than the usual Western vegetable grater with sharp-edged holes; simply peel the root and rub it vigorously against the grain over the spikes. In lieu of an Oriental grater, the food processor is your best choice. If the processed ginger is too fibrous for your purposes, spoon it into a piece of cheesecloth and twist to squeeze out the juice. For small quantities, a garlic press is just the gadget to squeeze the juice and nonfibrous pulp out of a bit of peeled ginger. Jars of grated ginger root, mixed with oil and other spices, are available, but the flavor is different and lacks the lovely brightness of fresh ginger.

dried ginger: Only relatively recently has fresh ginger been reliably available in the Western supermarket. Until this time, ginger was imported in in dried pieces, and of course it can still be found in this form. This ginger is dried at the point of origin, either in the sun or over low artificial heat. It may be peeled or unpeeled —also termed "white," "uncoated," and "scraped"; or "black," "coated," and "unscraped," respectively. Peeled dried ginger root is sometimes also bleached. Spice companies grind dried ginger to a fine powder, but it does not take commercial equipment to grind the dried hands, as they are quite friable, and you can easily do this yourself.

Note that dried and fresh ginger have different tastes, and one is not properly substituted for the other. Almost all Asian recipes want fresh ginger root, while traditional European and American recipes that call for ginger are referring to dried ginger.

Powdered dried ginger, properly kept, is preferable for breads, cakes, and other baked goods. A light sprinkling over melon or fruit cups intensifies their flavors. Ginger's warm, sweet flavor is particularly compatible with **oranges** and orange zest. Ginger also goes well with carrots, pumpkins, and parsnips, whether in a sweet or savory dish.

Middle Eastern cooks do not use a lot of ginger, but dried ginger is also the appropriate form of the spice in this cuisine; it is occasionally used with fish, chicken, and lamb or kid (especially in Morocco), as well as in a few sweets. Indian cooks, appreciating the virtues of both fresh and dried ginger, occasionally call for both in a single recipe.

Ginger appears in many spice blends around the world, including Ethiopian **berbere**, Indian **chaat masala**, Moroccan **ras el hanout**, and American pumpkin pie spice. It is also important in most **curry powders** and in **pickling spice**.

Popular ginger beverages include nonalcoholic ginger ale and ginger beer. There is not a well-defined difference between these two drinks; both are sweet and carbonated, but you might expect some versions of ginger beer to be spicier. "Ginger ale" tends to be the American term, while "ginger beer" is used more in Britain, where it may contain some alcohol. A refreshing homemade "gingerade" is easily concocted from a sugar syrup in which disks of fresh ginger have been boiled: mix the syrup with still or sparkling water and pour over ice. Ginger liqueur is not only a good after-dinner digestive: it is useful in the kitchen, too. Sprinkle a little ginger liqueur over fruits, as well as carrots and other sweet vegetables; stir a tablespoon into whipped cream and dessert sauces. The old-fashioned practice of sprinkling powdered ginger on top of a draft of regular beer is still worth trying, but you might have to skip the subsequent step in the traditional ritual: stirring the spice into the drink with a hot poker.

candied ginger: Fresh, young, tender pieces of ginger can be cooked in sugar syrup until candied. They may be eaten just as they are, as a near-perfect snack, or served with buttered toast or other breads for afternoon tea. Cubes of mild candied ginger, with the syrup drained off, may be dipped in melted **chocolate** to create a wonderful, electric confection. The spiced syrup, with or without the chunks of ginger, makes an excellent sauce for desserts such as ice cream or apple cake.

"Crystallized ginger" is candied ginger without the syrup, which has been rolled in granulated sugar to dry it. These crystallized pieces can be chopped fine and sprinkled on as a topping for *crème brûlée* and other desserts; they also make an interesting fruitcake when included with the candied fruits and peels.

All candied ginger is preserved by the action of the sugar, but the term "preserved ginger" has acquired a special meaning; it often indicates a candied ginger product that includes **licorice**, **salt**, and other spices as well. Preserved, crystallized, and candied ginger all last indefinitely.

pickled ginger: Ultra-thin slices of fresh ginger root, salted and pickled in sweet rice wine vinegar, make a spicy-hot condiment that clears the palate between flavors. The slices naturally develop a delicate pink tinge, but in Japan this is often accentuated with red **perilla** leaves and/or with red *omeboshi* vinegar (see Flavored Vinegars in the "Flavor Combinations" chapter). Pickled ginger is commonly eaten with sushi. It is also good with grilled meats of all kinds and with the stronger flavored fish such as tuna or salmon. Mix a few small pieces in with hot fried rice or a cold rice-and-vegetable salad.

To make an attractive presentation when serving pickled ginger, cut the slices in half and stand them on the cut edge, curling them around each other like the petals of a rose.

oil of ginger and ginger extract: While ginger's liquid form is convenient, these preparations have a different flavor than the juice of ginger obtained by grating or squeezing the fresh root. The oil tends to be bitter, so use it with some kind of sweetening.

myoga ginger (*Zingiber mioga*): This species of ginger is culti-vated in Japan, not for its rhizome but for its green shoots in spring-time. They are occasionally pickled, but usually eaten raw, often as a garnish for *sashimi*.

golpar (*Heracleum persicum*): There is no good English name for this plant, whose seeds are a popular seasoning in Iran. Many cook-books on Persian cuisine translate it as "angelica seed," and indeed it belongs to the same family of plants as **angelica**, the Apiaceae, and grows with a similar umbel of flowers producing small fruits generally called seeds. Several years ago, it used to be dubbed "marjoram seed" in English-language cookery books and in Persian food shops, but that is farther from the mark. It would be more accurate to call it a kind of cow parsnip; however, there are many different kinds of cow parsnip, not all of them good to eat. Since the

Iranians long ago gave this spice a perfectly good name, easy to say, simple to write, let us join them in calling it golpar.

Golpar is usually sold in powdered form. However, whole seeds are occasionally available; these are flat, pointed ovals with two dark spots at the rounded end; they are distinctly aromatic with a clean, balsamic smell. If you can find it, the silver variety of the plant is considered superior. The whole seeds should be ground up before using.

This spice is most often used in making all kinds of vegetable pickles, an ideal setting for its wonderfully sharp, almost hot, taste that lingers in the mouth. The powder is also sprinkled over beans and other pulses, as well as potatoes; it is also cooked in soups and stews to accompany rice. Golpar appears in many of the Persian spice mixes called **advieh**. The spice is used mainly in the northern parts of Iran in the Caspian region, and much less in the south of the country.

grain of Paradise/malagueta pepper/Guinea pepper/Guinea grain (*Aframomum melegueta*): There is some confusion about what this spice is, both because it is unfamiliar to us today and because of its several different names. "Guinea" refers to the country of West Africa, in the area where the spice is grown, while "malagueta" (probably) derives from the Spanish city of Malaga, where there was a thriving trade in this peppery spice centuries ago. The word "grain" is taken in its old meaning of "a small, hard seed," and whether the reference to Paradise is warranted or not is up to you to decide.

These small, red-brown, pointed grains are found densely packed in the fibrous pulp inside an elongated red seedpod. In West Africa, the pulp around the seeds is eaten, and the seeds themselves have many medicinal as well as culinary uses. They are an inevitable component of the Moroccan spice mix **ras el hanout**. During the Middle Ages, they were favored in Europe as one of the warm spices, along with **cinnamon**, **cloves**, and **ginger**. Many medieval recipes call for all of these spices together to flavor wine and sauces. Grains appear in numerous others of medieval Europe's

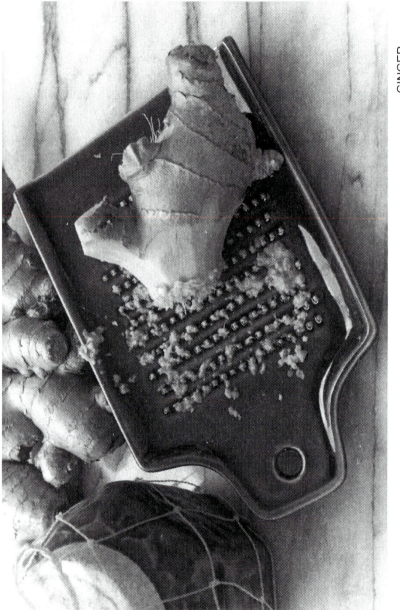

richly seasoned dishes. (Note that many translators of medieval texts have identified grains of Paradise as **cardamom** seeds: While both cardamom and grains of Paradise come from plants in the ginger family, and bear a superficial resemblance to each other, they are not the same and their tastes are very different.)

Grains of Paradise are biting on the tongue, like black **pepper**. They are used in the same way as pepper, either pounded in a mortar or ground in a pepper mill. Grinding releases their sweet aroma, and exposes their white interior. Black pepper is the most suitable substitute for grains of paradise, although **chiles** have sometimes stepped in. Today in Brazil, "malegueta pepper"—an essential ingredient in the national dish *feijoada*—is a particularly hot little chile.

Your best chance of finding grains of paradise is among the bulk herbs and spices at a health-food store.

H **hoja santa/hierba santa/acuyo/Mexican pepper** (*Piper auritum*): The large heart-shaped leaves of this tall Central American herb are not yet available in the produce sections of supermarkets across the country, but this may well change as Americans continue to embrace ethnic cuisines and seasonings of all kinds. Hoja santa will have to be sold as a fresh herb, because the leaves are not used dried. It is easy to cook with, and its pleasant **anise** flavor with herbal, minty overtones is easy to like. The aroma carries enough of a whiff of black **pepper** to remind you that the two seasonings are closely related, belonging to the same genus.

Hoja santa grows well in southern and eastern Mexico and in the warmest states of the United States. This attractive plant holds its large leaves horizontally around one or more thick central stalks. The leaves are easily six inches—often a foot—across, bright green on top and paler underneath. They are usually used as wrappers for steamed or baked fish, shrimp, chicken, or cheese. Or they may serve as an inner wrapper for tamales. Unlike banana leaves or corn husks, these leafy wrappers are eaten right along with the filling.

To make a wrapper, rinse each leaf well, lay it on a board and slice along the central vein on both sides, cutting the heart shape into two lobes. Discard the tough central vein and use the two large pieces as wrappers. You can also make a chiffonade of them (see the "Culinary Practice" chapter) or use them as a seasoning.

Because they are tough, hoja santa leaves are not good in salads. They need to be cooked, but the good news is, they keep their flavor and remain green when heated. They are used in Mexico to season a mole for pork and are sometimes added to *posole verde*. For these purposes, you may substitute the feathery green leaves of **fennel** for hoja santa, if necessary, using about one-half cup fennel leaves for each hoja santa leaf called for in the recipe.

With a blender, it is possible to make a hoja santa sauce, good with fish or chicken. Tear up a few leaves into largish pieces, and pack them loosely into a measuring cup. As you tear, you may find a few more tough veins to discard. Put one cup of leaves and one-third cup chicken broth into the blender and puree as smooth as possible. Season with salt and pepper, and with **garlic**, onions, green **chiles**, or whatever suits your mood. Now fry the sauce, as they do in Mexican cooking: pour into a skillet with a small amount of oil and bring to a boil, stirring frequently. Turn down the heat immediately, simmer for about ten minutes and serve. This very green sauce looks wonderful with an accent of finely shredded carrots.

The leaves can be cut into strips and fried crisp, as with fried parsley. Fry them in hot oil for two minutes or so, until they turn dark green and curl up. Drain on a paper towel, and add a pinch of salt. Use as a garnish for any entree, especially fish, or with vegetables such as squash.

Hoja santa tastes so good with fish that in parts of Panama the leaves are fed to live, stocked fish, which then acquire the flavor of the herb.

hyssop (*Hyssopus officinalis*): The narrow, dark green leaves of hyssop have a minty aroma and a strong, bitter taste that is penetrating and persistent. The small purple-blue flowers—occasionally appearing in pink or white—have the same flavor but are somewhat milder. Hyssop is a favorite of the makers of bitters, herbal liqueurs, and digestives, but many people find it too pungent to use much in cooking.

Use this herb sparingly; it can easily overpower the other flavors in a dish. Only two or three leaves or flowers suffice for an individual serving of green salad. But a little hyssop can be a pleasant surprise in simple fruit dishes such as compotes or stewed prunes. It will add a great deal of interest to a peach cobbler, which might otherwise be sweet and mellow but not much more. Just sprinkle a scant teaspoon of ground dried hyssop, or twice as much of the finely chopped fresh leaves, under the crust. A pinch of the herb is good in pea soup and in lentil and mushroom dishes.

Sausages, pâtés, and meaty stews are often seasoned with this robust herb. A little hyssop will anchor the flavor of a savory fruit sauce for duck, goose, or turkey.

In Israel, hyssop is sometimes used in place of **thyme** to make **za'atar**. It is debated whether today's hyssop is the same as the herb mentioned by that name in the Bible, and the matter is unlikely to be settled definitively.

The American plant called anise hyssop, *Agastache foeniculum*, is botanically unrelated to hyssop. Its pointed leaves and purple flowers smell and taste of **anise** and produce a flavorful tisane (see Teas and Tisanes in the "Culinary Practice" chapter). They may be used fresh or dried in place of anise seed in any recipe.

J **juniper berry** (*Juniperus communis*): The prickly juniper shrub grows sedately over most of the northern temperate zone. Its fruits require two to three years to ripen, and a single female bush usually has berries in various stages, ranging in color from a juvenile light green to a mature blue-black. Gloves are needed to harvest the berries, because the juniper needles are extremely sharp.

Most juniper berries on the market are harvested in southern Europe, where their flavor is more potent than that of berries grown in colder regions. Nevertheless, this spice is most popular in the cooking of northern Europe, especially Sweden, Poland, and Germany.

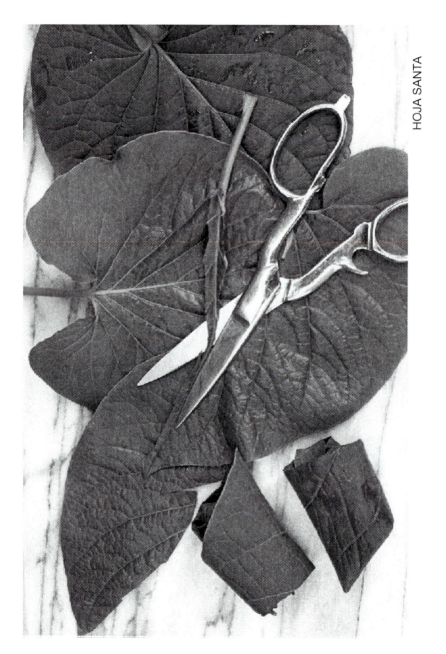

The berries are usually dried whole and sold in this form. Stored in the usual cool, dry, dark place in an airtight container, they will keep almost forever. In fact, you have to help them release their flavor when you cook with them by crushing them lightly with the back of a fork.

The flavor of juniper berries is bright, sharp, and slightly resinous—the taste, in fact, of gin. Our word "gin" comes from the Dutch *genever*, meaning "juniper." If you are out of juniper berries, you can substitute gin in most recipes; cooking will drive off the alcohol, leaving only the juniper flavor. Use one-fourth cup of gin to replace six to eight berries.

Juniper berries are added to marinades for game, venison, pork, oxtail, duck, and other strong-flavored meats. Red wine or red wine vinegar provides a good basis for a juniper marinade, and **bay leaves**, **garlic**, and **parsley** are compatible seasonings. Such a combination makes a good braising liquid for these meats as well. Just a few berries are needed: four to six per pound of meat is a rough guide, but juniper berries do vary in size and intensity of flavor.

Juniper berries also make an exciting sauce for these meats. Simply melt a little apple or grape jelly with lightly crushed berries, and allow it to cool and gel again. In this case, restraint with the berries is not necessary: the amount of juniper flavor will be determined by each eater who takes a dab of sauce with the meat.

Juniper berries are also good in many a pâté or terrine. They are a popular seasoning for sauerkraut and red cabbage. Try adding a few to mincemeat or to fruitcake. The berries add excellent flavor to stewed rhubarb.

The question arises, how much should the berries be crushed? Try pressing with the fork just enough to split the skin on one side, exposing a glimpse of the interior brown pulp. This produces relatively whole berries that are very attractive in a dish, but which do provide a flavor jolt when you bite into them. You may welcome this savory salvo, or you may prefer to pick the berries out after they have done their job of infusing flavor into the dish.

At the other extreme, some people like to pound up the berries finely with a mortar and pestle. A pinch of sugar or salt in the mortar helps to keep them from sticking. This creates a more strongly flavored dish, but the little seeds nestled in the pulp provide a crunchiness that may not be to everyone's taste. Another disadvantage of this technique is that it requires you to know your culinary audience and their taste for juniper berries, whereas with whole berries diners have the option of enthusiastically gobbling them up or discreetly pushing them away to the edge of the plate.

The fresh clean taste of juniper berries resembles that of **angelica**, and these two seasonings can be used together or substituted for each other.

L **lavender** (*Lavandula angustifolia*): The name "lavender" is probably derived from the Latin *lavare*, "to wash," indicating the ancient use of the plants. But the bright, optimistic scent of lavender is as welcome in the kitchen as in the laundry or the bath. Of course, any lavender for culinary use must be organically grown, free of all pesticides and other toxins.

Of the many species of lavender, the favorite for cooking is English lavender, *L. angustifolia*. Other species can also adapt to the kitchen, especially *L. stoechas*—known as Spanish, Italian, or sometimes French lavender—and some varieties of the hybrid lavandin, *L.* x *intermedia*. A little experimentation with these plants will reveal the best way to use them.

Both the leaves and flowers are edible, and they may be used fresh or dried. In practical terms, what you're most likely to get is dried flowers, generally available in bulk at health food stores. Even if you have lavender in your garden, the short blooming period makes it necessary to use dried flowers most of the time. Lavender leaves, on the other hand, are present year-round on these evergreen shrubs and may be used in lieu of the flowers if you like them, although their flavor is a little harsher, not quite as bright and sweet as that of the flowers. (More on cooking with lavender leaves follows.) The woody stems can be placed on the coals when grilling fish or poultry to add an intriguing flavor and to perfume the air. Oil of lavender is fiercely potent and not recommended for internal use.

Both flowers and leaves are inclined to be tough, and the trick in cooking with lavender is to transfer their fragrance and taste to the food, while the plant parts themselves are generally discarded. The tenderest—and, incidentally, the sweetest—parts of the plant are the colorful petals of fresh flowers. Carefully plucked from the hard base of the flower, the delicate petals can be scattered over cakes and other desserts to great effect. These treasures can be preserved past the blooming season with a thin coating of beaten egg white and fine sugar; to avoid all risk of salmonella, use powdered egg white or a solution of gum arabic and sugar. Handle the petals with tweezers, be certain they have completely dried after coating, then store in an airtight container.

The simplest, most obvious way to extract flavor from this plant is to brew a lavender tea (see Teas and Tisanes in the "Culinary Practice" chapter). Lavender is potent; you will probably find half a teaspoon of dried flowers sufficient for a cup of tea. Strain out the flowers, and drink the hot tea with honey to soothe the nerves and delight the senses.

Steeping lavender in warm cream or milk, then straining out the flowers, leaves a delicate, scented liquid for making lavender custard, ice cream, and sauces.

Similarly, burying dried flowers in sugar for several days transforms that sweetener into an elegant flavoring for shortbread, cookies, or lavender meringues. The lavender flowers may be stored in the sugar indefinitely, but should be sieved out before the sugar is used.

The above methods leave a light breath of lavender in the cream or sugar. If you're in a hurry, or simply want to emphasize the presence of lavender, there are ways to speed up and intensify this transfer of flavor. But before you try any of the following tips, note first that the taste of lavender is quite sharp and can definitely be overdone; consider increasing only the aroma component of its flavor by garnishing the dish with a beautiful sprig of dried flowers or a stem of scented fresh leaves (see Flavor in the Introduction).

To add more lavender flavor, use both lavender sugar and lavender-scented cream in a single dish. Or, if the recipe is amenable to using a syrup, heat the flowers with a mix of water and sugar, and strain out the lavender while the syrup is still warm. Or buzz granulated sugar and a little dried lavender in an electric mill and push it through a sieve to make fine powder that can be used like confectioner's sugar to dust the tops of cookies, tea cakes, and so on. The minute lavender-colored flower bits in the sugar are fine enough to be pleasant to eat and they are as pretty as they are fragrant and tasty. This lavender-sugar powder is also good with whipped cream or yogurt.

For more robust dishes, dried lavender flowers are simply pulverized in an electric grinder, then sifted directly into the food. The latter method is standard when cooking with **herbes de Provence**— a warm, earthy blend of herbs that often includes lavender.

The flavor of lavender blends well with citrus fruits of all kinds. Try sprinkling a little lavender sugar over a grapefruit half. Or enrich the flavor of lavender tea with a splash of **orange** juice. Lavender sugar takes orange marmalade to a new level: simply heat a store-bought marmalade, set the peel aside temporarily, then add a spoonful of the flowers and let the warm mixture sit for about five or ten minutes. Finally, strain out the lavender and return the peel before the marmalade gels again. For another sensual citrus dish, make ambrosia with lavender sugar. Or use lavender-scented milk to make scones studded with bits of candied citrus peel.

Lavender also enjoys a tryst with **chocolate**. Try lavender sugar in your hot chocolate, or a dark chocolate syrup on your lavender ice cream. Ice a lavender-flavored cookie with chocolate frosting, or top a chocolate cake with lavender cream, or melt chocolate in a lavender cream to make a romantic pudding, or . . . you get the idea.

Lavender flowers may infuse an herbal vinegar, which is ideal for salad dressings and for marinades.

Bees love lavender, and make a gourmet honey of its nectar. Another kind of lavender honey can be made by steeping the flowers in any mild honey for a week or longer. Warm the honey to make it flow better when you want to strain out the flowers. Either type of lavender honey is excellent on tender scones, flaky biscuits, or hot buttered toast.

Lavender does not give up more than a tinge of its color with its flavor. If you dare, you may devise a lavender tint for lavender cream or ice cream with two drops of blue and one drop of red food coloring (don't use it all; start with one drop of the coloring mixture), but most diners are happier with the look of a few fresh or candied petals or perhaps a sprig of flowers laid on the edge of the plate.

Lavender leaves resemble **rosemary** in that both herbs have penetrating and persistent, somewhat piney flavors. Chopped lavender leaves can substitute for rosemary in herb breads, or in a compound butter for grilled fish, or in a bread stuffing for poultry. The lavender is usually milder than the rosemary—a fact that may be welcomed by those who object to rosemary's assertiveness; adjust the quantities of lavender according to your own preferences.

Lavender leaves make sensual garnishes for all kinds of foods. Consider a sprig beside a slice of lemon cake, or a bed of leaves for roast lamb, or a platter of roast chicken encircled by a garland of lavender leaves and lemon wedges.

lemon (*Citrus limon*): Lemons are indescribably important as a flavoring agent. While they are not very aromatic, they have a no-holds-barred taste. The fruits can be used in several ways: lemon juice, lemon leaves, and lemon zest all differ in intensity, freshness, and oiliness. Be sure to read the Citrus section of the "Culinary Practice" chapter for many tips on seasoning with this popular fruit.

Where the French generally use vinegar, Middle Eastern cooks prefer fresh lemon juice. Mixed with olive oil, it makes a refreshing, fruity dressing for Greek salads. Rice-stuffed green peppers, zucchini, tomatoes, and vine leaves are soaked in this mixture and served at room temperature. (In contrast, meat-stuffed vegetables are served hot and without the oil and lemon.) Thin slices of lemon, sealed in a pot with chunks of lamb and sweet onions, make a simple savory stew, light enough for any season; top it with chopped flat-leaf parsley for color. *Avgolemono*, a rich egg-and-lemon soup, is popular in Greece on festive days.

Lemon and **mint** are always a good combination. The pair stars in "Saudi champagne," a refreshing nonalcoholic drink made with equal parts of apple juice and sparkling water, brightened with lots of lemon slices and generous amounts of fresh mint sprigs. Serve over ice.

Nature has produced a number of plants with lemony tastes of varying intensity, including **balm**, **lemon grass**, lemon **thyme**, **lemon verbena**, and lemon-**scented geranium**. While these plants have lemon's flavor, they do not have its acidity, so they may be added to milk without curdling it. Conversely, citric acid, or "lemon salt," has the acidity of lemons without the flavor. See the Citrus section of the "Culinary Practice" chapter for more information about this essential acid.

preserved lemon: Another uniquely delicious flavor can be obtained from the lemon by preserving it in salt, as is done regularly in Morocco and other parts of North Africa. The whole fruit sits in salty juice until the peel is translucent, soft, and tangy but not sour.

Preserved lemons are easy to make: Sterilize a large glass jar in boiling water for about twenty minutes. The jar should have a lid of glass, plastic, or cork; a metal lid would be corroded by the salt. Meanwhile, wash thoroughly a quantity of whole lemons, just one or two more than enough to be squeezed into the jar. Organic lemons are recommended, especially since the peel is the part of the fruit to be used.

Now comes the fun part: Make a long slice from one end longitudinally through the center of the lemon to about one-half inch from the opposite end. You want the fruit to hold together. Then turn the lemon over and do the same thing from the other end, slicing at right angles to the first cut. Thus the fruit is *almost* cut into four long wedges, but it remains in one piece; a large part of the interior is accessible for salting.

Fill each slit generously with a teaspoon of sea **salt** or kosher salt—not table salt, which contains additives—and press the fruit together again. Put the salted lemons into the jar, squeezing in as many as possible, and cover tightly. Leave on the kitchen counter overnight.

The next day, you should be able to squeeze in another lemon or two, salted in the same way. If the juice that has collected in the jar does not cover the lemons, add some fresh-squeezed lemon juice and a little additional salt. Then move the sealed jar to the refrigerator for about one month.

The peel of these preserved lemons, cut into thick strips and laid on top of a dish, makes a garnish as attractive as it is edible, and it is essential to many Moroccan dishes such as chicken tajine with lemon and green olives. Although it is the peel that is most important, a little of the lemon pulp can also be used in cooking, and the juice of preserved lemons is excellent as a substitute for vinegar in homemade salad dressings.

lemon grass (*Cymbopogon citratus*): This popular Southeast Asian plant looks very much like tall grass, and indeed it is a member of the grass family, Gramineae. The leaves are long green blades with slicing-sharp edges; several of them grow up from a single white fleshy stalk that resembles a green onion's. This seasoning has recently become so popular that fresh lemon grass has begun to appear in supermarkets. If you live in a sunny climate, however, you might grow it yourself, either buying a potted plant from your nursery or rooting a fresh stalk in a glass of water.

In Latin America, a lemon grass tea is made from the leaves. Keep the leaves whole and, being careful of their sharp edges, fold a large handful of them into a package small enough to fit in a saucepan. Fill the pan with water, bring it to a boil, and let it boil gently over medium heat for ten to fifteen minutes; don't overcook, or the tea could turn cloudy. Remove the leaves and serve the winsome golden liquid hot or cold, sweetened or unsweetened, as you desire. If the tea is meant to be iced, then make it extra strong with more lemon grass. This is as refreshing a beverage as you could ever thirst for.

Traditionally, this flavorful tea is the only use of lemon grass in Latin America. The stalk is not used; it will be left in the ground when the leaves are cut, to grow a new supply. On the other hand, in Southeast Asia—from Thailand through Cambodia, Laos, and Vietnam to Malaysia and Indonesia—there are innumerable recipes that use lemon grass. In these countries, only the stalk is used, and the leaves are discarded. With lemon grass, you can be both Jack Sprat and his wife, keeping the stalk for later use when you make lemon grass tea, or setting aside the leaves when you use the stalk in a Southeast Asian dish. Both leaves and stalk can be stored in the refrigerator for two or three weeks, and will keep even longer when frozen. The fresh leaves can also be dried and kept for several months.

Most of the stalk is much too fibrous to be eaten. It is usually cut into large pieces and allowed to flavor the dish, then strained out before serving. Southeast Asian cooks sometimes bruise the stalk to release its oils, tie it in a loose knot to keep the fibers from escaping, then throw it into the pot whole for easy removal when the dish is finished. Only the tenderest inner part of the root end of the stalk can be eaten, and it should be finely chopped or very thinly sliced.

This versatile flavor is especially welcome with beef, fish, seafood, and poultry. Simply boiling a skinless chicken in water with two stalks of lemon grass creates a savory meat to serve cold in a salad. The resulting broth can be incorporated into a soup; decorate each bowlful by floating in it two or three ultra-thin disks of tender fresh stalk. Often the juicy root end of the stalk is pounded into a paste with other spices, to season stir-fries or such dishes as Thai curries, Vietnamese marinated beef, or Indonesian spiced coconut milk with crabs.

Lemon grass has a decided affinity for hot **chiles**, as anyone who has eaten Thai food knows. It also marries well with **coconut** milk. A dilute lemon grass infusion can be used to make a delicately flavored rice, to serve with fish or shellfish, or to make a cold rice salad.

The subtle perfume of lemon grass is excellent in desserts as well. Lemon grass sherbet is hauntingly good, especially after a spicy hot meal. You can make a *crème brûlée* or a basic cream sauce with lemon grass: omit the usual vanilla flavoring and let a stalk of lemon grass infuse the cream as you scald it. Leave the stalk in the cream as it cools; strain it out just before the cream is called for in the recipe. For a tropical flavor variation, substitute coconut milk for cream. Candied lemon peel makes a compatible decoration.

Dried lemon grass is sold in shreds or in powder form, the latter sometimes labeled "sereh powder," but this is so inferior to fresh lemon grass that you may be happier with the fresh substitutes recommended below. If you do use the dried forms of lemon grass, measure three tablespoons of dried shreds or one tablespoon of powder to replace each fresh stalk. Soak the dried herb in warm water for at least half an hour before using; then strain out the bits of lemon grass stalk and chop them very fine.

Because lemon grass is lemony but not sour, lemon *juice* will not substitute for lemon grass, but the *zest* of the citrus fruit, which contains citral, the same aromatic essence that is in lemon grass, can be used. Substitute chopped lemon zest for chopped fresh lemon grass in equal quantities. Or simmer the zest of one medium-sized lemon in place of each whole stalk of fresh lemon grass. The zest of a large **lime** (not a key lime) may also be used. The addition of one-quarter-teaspoon of grated fresh **ginger** will improve all these substitutions; in fact, if no lemon or lime zest is available, a different but similarly bright flavor can be achieved with ginger root; use about one teaspoon of grated fresh ginger in place of a stalk of fresh lemon grass.

lemon verbena (*Aloysia triphylla*): This South American native is popular for its intense perfume and its delicate taste, which is both lemony and herbal. Dried lemon verbena leaves are not very difficult to find—natural food stores often sell them in bulk—but the fresh leaves are elusive. You might grow your own. The plant is a pretty, slow-growing shrub that must be kept warm in the winter. It's happy to be kept in a pot, so you can move it indoors when necessary.

The classic use of lemon verbena is in a tasty tisane (see Teas and Tisanes in the "Culinary Practice" chapter). Use a rounded teaspoon of dried leaves or a tablespoon of chopped fresh leaves for one cup of hot water; then adjust the proportions to your taste for the next cup. This tea is especially mellow when sweetened with a mild honey.

A similar infusion of lemon verbena leaves in vinegar (see Flavored Vinegars in the "Flavor Combinations" chapter) makes a good basis for salad dressings and marinades for poultry.

Finely chopped fresh leaves are excellent in stuffing for chicken or turkey, and in an herb butter for poultry, fish filets, tuna steaks, or crabmeat (see Compound Butters in the "Flavor Combinations" chapter). If you cannot find the fresh leaves, crumble up dried ones and remove the larger pieces of stem.

Sprigs of the fresh leaves make an aromatic garnish for fruit salads and frozen fruit sherbets, or they can top off an icy pitcher of tea, lemonade, or punch.

When necessary, any one of the lemon-scented herbs—lemon **balm**, lemon **thyme**, lemon-**scented geranium**, **lemon grass**, and **lemon verbena**—will make a tolerable substitute for any other. But their flavors are not identical. Lemon verbena is the sweetest of these, and its dried leaves hold their scent best. None of these will replace lemon juice, because they lack the essential acidity, but in a pinch you might try replacing lemon zest with one of these herbs.

licorice/liquorice (*Glycyrrhiza glabra*): A native of the Mediterranean region, this legume has been cultivated over a very wide range in Europe and Asia. The plant has been prized since antiquity for its roots, which have a pleasing flavor and potent medicinal effects. Please note that you should not use this powerful herb regularly, especially if you have hypertension or problems with water retention.

A close look at the plant's generic name, *Glycyrrhiza*, reveals the origin of our word "licorice"; the name has nothing to do with liquor, in spite of the variant spelling. *Glycyrrhiza* means "sweet root," and indeed one component of licorice root, called glycyrrhizin or glycyrrhizic acid, is astonishingly sweet. At one time it was thought to be the sweetest natural substance on earth; but that was before much was known about the South American herb stevia or sweet herb of Paraguay (*Stevia rebaudiana*), which contains an even sweeter compound.

Licorice root also contains anethole in its essential oil, which figures so prominently in the flavor of **anise** seed and is also present, in varying degrees, in **avocado leaf**, some **basil, chervil, fennel, hoja santa**, Mexican mint **marigold, star anise**, and **tarragon**, as well as wormwood, which used to flavor the now-banned liqueur absinthe. (While anethole gives a shared characteristic to all these seasonings, they certainly do not all taste the same; each one has its own unique flavor.)

Licorice roots look like thin sticks, brown on the outside, with a yellow wood inside; they are usually cut to lengths of four to eight inches. The best places to find licorice sticks are in the bulk-herb jars at health food stores and at ethnic shops among the herbs offered for tisanes. Chop the roots coarsely for a tea, or buzz the pieces in an electric coffee mill, then sift out the woody bits to get a fine powder. Sometimes you can find licorice already ground.

To make a simple licorice tea, add pieces of coarsely chopped or powdered root to a pot of cold water—one scant teaspoon chopped root per cup of water is sufficient—and bring to a full boil. It will foam. Take the pot off the heat, cover, and let the infusion steep for five to ten minutes, or longer if you want your tea stronger. Then strain it and enjoy; no sugar or honey is necessary because of the natural sweetness of the herb.

Street vendors in nineteenth-century Paris sold a drink made from licorice and **lemon**, calling it "coco" because of a supposed resemblance to exotic **coconut** milk. In fact, adding lemon juice to a licorice infusion produces a slightly cloudy mixture with a grassy yellow tinge, nothing like coconut milk, but looking and tasting like a rather full-bodied lemonade. Under any name, however, this simple drink is delicious, refreshing, and satisfying. It is easy to make: adjust the proportions so that the acid lemon juice is balanced by the delicate sweetness of the licorice tea. The beverage is sensational when iced.

Chewing on the sticklike licorice roots has been popular down the centuries, both for the enjoyable touch of sweetness and for their thirst-quenching qualities. It was a short step onward to making licorice candy in the form of sticks of about the same diameter and length. When the juice extracted from licorice roots is reduced, it becomes black, sticky, and sweet, and so these candy licorice sticks are traditionally black. The image has been immortalized in the jargon of jazz, which labels the clarinet a "licorice stick."

This substance obtained from boiling down licorice juice, sometimes called "black sugar," is the traditional basis for licorice cough drops, syrups, and candies in an endless assortment of shapes and combinations with other flavors. Dutch confectioners produce both sweet and salty versions of licorice drops for an eager population. Some candy makers augment the anethole flavor component of licorice by the addition of a little oil of anise—sometimes to the point of omitting the licorice. Genuine licorice candy is far more popular in Europe than in America.

Almost all of the licorice root imported into the United States is used by the tobacco industry to sweeten its products; a certain amount is also included in cough syrups and other medicinal products; and some licorice is used in brewing beer, especially dark beer, to which it contributes color, flavor, and foaming action.

We have gingerbread recipes from the late Middle Ages that call for powdered licorice root as part of the spicing. This is an excellent flavor for that venerable baked good and, if you're fond of licorice, you might like to try adding it to your own gingerbread. Start with about a teaspoon of licorice powder for each cup of flour, and adjust your recipe according to taste. You will need to reduce the sugar (or honey or molasses) since licorice is so sweet: try removing two tablespoons of sugar for each teaspoon of licorice powder, and see if the recipe wants further adjustment.

Licorice is used in Chinese cooking in soups and to season red-cooked dishes—those luscious meats slow-braised until tender in multispiced soy sauce. Either a spoonful of powder or dried slices of root are added to the liquid. Large slices are then removed before the dish is served; smaller pieces may be strained out or left in the broth. A little ground licorice root sometimes appears in Chinese **five-spice powder** to enhance that anethole flavor. The Chinese use the species *Glycyrrhiza uralensis*, as well as the European *G. glabra* and one or two other licorice species.

Anise seed is the usual stand-in for licorice, although the feminine, anethole flavor of anise is not the same as the sweet, rooty, more masculine taste of licorice.

lime (*Citrus aurantifolia*): See also the Citrus section of the "Culinary Practice" chapter. The most common variety of lime used in North America is the so-called Persian lime, which looks like a small, bright green lemon. Limes are tarter than lemons and have a more piercing flavor, with a hint of bitterness that is missing in lemons, but they have a sweeter, headier perfume. There is a distinctive tropical tang in the taste of lime.

Lime flavor marries well with other strong flavors: Just as **lemon** and **parsley** are a perfect match, lime and **coriander** leaf make a blissful pair. A compound butter of lime juice and finely chopped cilantro is perfect on grilled fillets of strong-flavored, oily fish such as tuna or salmon.

Lemons and the standard Persian limes may be substituted for each other when necessary.

Some languages such as Arabic, Persian, Spanish, and Turkish do not distinguish between lemons and limes in the same way that English does, but use a single word—generally sounding something like "li-moon"—for both fruits. When necessary, distinctions are made between them and their many varieties by means of adjectives and other descriptors. This situation can lead to confusion in cookbook translations, so when reading cookbooks from other cuisines, keep in mind which fruit is more likely to be used in the region you're dealing with. To oversimplify the matter, limes generally grow in more tropical climates than lemons and are generally preferred to lemons in the cuisines of warmer places, such as Mexico, the Caribbean, parts of Africa, India, and Thailand; while lemons are found much more often in the foods of temperate regions such as Europe, most of North America, and the Middle East.

South of the (U.S.) border, wedges of fresh limes are often served with tropical fruits, such as papayas, various melons, and jicama, to be squeezed over them as a bright flavor accent. In Mexican and South American *seviche*, raw fish is "cooked" in acidic lime juice before it is finished in a piquant sauce.

Limes are very good with many alcoholic drinks from margaritas to planter's punch and other rum drinks to Mexican beers in which a lime wedge is served in the neck of the bottle, inviting the imbiber to push it down inside. Use lime juice to make coarse **salt** crystals adhere to the rims of margarita glasses.

In Southeast Asia, lime juice garnishes soups, salads, and noodle dishes, and a few drops of lime juice go into *pho*, those abundant soups so loved by the Vietnamese. Often wedges are served so that each diner can add juice according to individual taste. Thai cooks make a salt-cured lime very similar to the Moroccan preserved lemon (see **lemon**), except that a little sugar is usually added to counteract the lime's greater tartness. These pickled limes make a most interesting condiment, especially well suited for spicy-hot dishes from any cuisine.

In general, lime is good with Oriental flavors. (See the **ginger** entry for a seasoning suggestion involving that spice with soy sauce and lime juice, p. 125.)

Lime and **chiles** make an exciting flavor combination. Some food companies have capitalized on that fact by offering tortilla

chips in a chile-lime flavor. In India, a favorite condiment is lime
pickle, a mouthwatering combination of pieces of lime, lots of
ground **chile**, spices such as **mustard** and **fenugreek**, and vinegar.
Lime juice and chopped chiles mixed into *nam pla*, the Thai fish
sauce, make a popular sauce for plain rice.

Lime also combines well with other citrus flavors, the intriguing
lemon-lime flavor used in beverages and candies being the prime
example. Try seasoning a summer afternoon by altering the standard lemonade recipe to include both lemon and lime juices; sweet-
en to taste and enjoy your lemon-limeade! Serving both lemon and
lime wedges with a bowl of hearty soup, such as a black-bean soup,
allows the diner to create interesting flavor variations.

Brown spots on otherwise firm, heavy limes do not affect their
quality, but a yellowish tinge does indicate somewhat lower acidity.
This is not true, however, of key limes, which turn yellow as they
ripen.

key limes/West Indian limes/golden limes: These small, round
yellow limes are named for the Florida Keys, where they grow
happily wherever they find space left unoccupied by the burgeoning
condo crop; they also thrive all around the Caribbean. Key limes are
very tart—tart enough to stand up to a rich custard of egg yolks and
sweetened condensed milk, the basic stuff of the famous key lime
pie.

Key limes add extra zing to marinades, salad dressings, and
margaritas. A little tart juice in a basic shortbread makes it extraordinary. And while the **orange** is a good partner for dark **chocolate**, the key lime is the perfect match for white chocolate!

Thin-skinned and juicy—one to two tablespoons of juice per
small lime—key limes are expensive, but a worthwhile investment.
They also contain about half a dozen seeds each, so you should be
prepared to strain the juice before using it in a recipe.

Don't bother trying to replace tart, tangy key limes with lemons.
However, if necessary, you can substitute Persian limes for key
limes.

Bottled lime juice and bottled key lime juice are both available;
they are convenient and tasty, but if you decide to use them you
must accept the fact that they taste less bright and, simply, different
from the juice of fresh fruit.

dried limes/limou amani /omani lemons/loomi/noumi/basra/ limoun basra: Dried limes are a popular seasoning all around the Persian Gulf—in Iran, Iraq, and all the countries of the Arabian Peninsula. These small, shriveled fruits add a musty, sour bite to Persian soups and stews, such as the famous *ghormeh sabzi*, an intricate stew of meat, kidney or fava beans, and a plethora of herbs such as **fenugreek**, **parsley**, **coriander** leaf, **dill**, onions, and **chives**. In the Gulf states of the Arabian Peninsula, their flavor is appreciated with fish and shrimp. And they are indispensable in *kabsah*, the Saudi national dish, which is made of rice and tomatoes, spiced with dried limes, **cardamom**, **cinnamon**, **cumin**, and **pepper**, with your choice of meat and additional vegetables.

Dried limes are usually used whole: washed, pierced a couple of times with the point of a sharp knife, and tossed into the pot for a long simmering in the stew or broth. One or two limes will season a dish for six to eight people. They can be removed at the end of the cooking, but they are more commonly served with the dish. The lucky diner who gets a whole lime on his plate generally flattens it with a fork to squeeze out every drop of the scrumptious juice.

If there is little liquid in a dish, or if it has a shorter cooking time, the limes can be pulverized and sprinkled over the food. Sometimes a little dried lime powder is rubbed over the inside of a whole fish before it is baked or poached. A tasty compound butter for grilled fish or poultry is made by mixing half a teaspoon of the powder with a stick (four ounces) of butter (see Compound Butters in the "Flavor Combinations" chapter). You can sometimes buy dried lime powder in Middle Eastern food stores, or you can make your own: First open the dried limes and remove the seeds (which impart a bitter taste when ground), then grind the remainder of the fruits to a fine powder, using an electric coffee grinder or a mortar, pestle, and elbow grease.

Some dried limes are pale brown, while others are very dark—almost black; the dark limes are more sour, and they will darken the broth when cooked. If you have a choice, use the paler fruits for light-colored dishes.

Arab and Persian groceries generally carry dried limes, but you can dry your own, if you prefer. Boil the whole fruits vigorously in water for a few minutes, then dry them in a sunny or otherwise dry and warm place for several weeks until they turn brown and feel hollow; they should bear a strong resemblance to dark, wrinkled ping-pong balls.

A simple, yet invigorating drink is made of dried limes. Crack open a lime and put the pieces into a teapot. Pour in two cups of boiling water and let it steep for at least five minutes. Strain the pale liquid into teacups and serve with the sugar bowl. Bliss!

kaffir lime/magrut (*Citrus hystrix*): This species of lime is cultivated not for its flesh or juice, but for the aromatic oils in the leaves and in the zest of the fruit. The compound kaffir lime leaf—looking like two leaves strung together on a long stem—is an important ingredient in Southeast Asian cooking. Recipes sometimes call for "fragrant lime leaves."

Fresh leaves, if you can find them, may be kept frozen; sometimes you can buy them frozen in an ethnic supermarket. Dried leaves also work quite well if they have time to simmer in a hot liquid for a few minutes. Use one or two whole leaves per person. The leaves are not eaten and may be removed at the end of cooking, like a bay leaf.

Whole leaves typically season Thai seafood dishes such as a fish curry with **coconut** milk and **chiles**. Shredded or julienned leaves of kaffir lime are used in Thai and Vietnamese chicken dishes, both as a flavoring and sprinkled over them as a garnish. Indonesian cooks might bury a lime leaf in an elaborate fried-rice dish, while strips of leaves or a little grated rind often flavor the incendiary Indonesian relish *sambal badjak*.

Kaffir lime leaves yield a remarkable amount of essential oil with a floral perfume that resembles the herbal flavor of **lemon verbena** more than a sharp citrus taste. Indeed, the best substitute for a kaffir lime leaf is two teaspoons of chopped, fresh lemon verbena or one-half teaspoon of the dried leaves. In lieu of lemon verbena, you can use half a teaspoon of finely grated lime zest for each double kaffir lime leaf. Both the verbena and the zest are more bitter than the lime leaves, so you may want to add a pinch of sugar when making these substitutions. Any other citrus leaf may also be substituted for kaffir lime leaves, but they should be added in double quantities; that is, one double leaf of kaffir lime is approximately equivalent to four medium-sized orange or other citrus leaves.

LICORICE

151

The zest of the small, very knobbly kaffir limes is also used as a flavoring. Grate the rind of any other lime to replace kaffir lime zest, and use twice as much as the recipe calls for.

Do not confuse the citrus lime trees with the so-called lime tree, or linden tree, of Europe, *Tilia cordata*, whose dried flowers are used to make linden tea, also called lime tea or lime flower tea.

M **mace** (*Myristica fragrans*): Mace grows on the same tall, leafy tropical tree that produces **nutmeg**. When ripe, the pale yellow, peachlike fruit splits open while still hanging on the tree, dramatically revealing a shiny black nut wrapped in a brilliant red net called the aril. The kernel inside the hard black shell of the nut is nutmeg; the bright-colored net around the shell is mace.

The mace is carefully removed from the nut, flattened, and dried. Its color rapidly fades to pink, then yellow, and finally to golden-brown as the spice ages. Pieces of the brittle dried net, usually narrow strips about an inch long, are called "blades" of mace. Mace is sold either as blade mace or commercially ground to a fine powder.

Mace and nutmeg, the two spices produced by this fine tree, are similar in aroma and taste but they are really not the same. Given the way they grow, each tree naturally produces far less mace than nutmeg, and mace is the more expensive spice. Mace has a much more potent taste, and it sings out its presence in a dish. Mace is a diva, while nutmeg is happy to blend in with the chorus. Be careful not to overdo when seasoning with mace.

There is also a qualitative difference between the flavors of the two sister spices that leads many writers to describe mace as "more refined" than nutmeg. Mace is richer and warmer, with undertones of cloves, while the flavor of nutmeg is sweeter and lighter, yet rougher than mace.

The color of ground mace is golden-brown, much lighter than the chestnut-brown of ground nutmeg. West Indian mace tends to be yellower than the East Indian, which is more orange.

Mace is difficult to grind at home, so don't buy blade mace if you want to use it as a powder. Purchase small quantities at a time of commercially ground mace to have it as fresh as possible.

Ground mace is the signature spice of doughnuts and a favorite flavor for pound cakes; it lends a soft golden tint to these products. A little mace is good in banana bread and in **chocolate** cake. Cherry pies become intriguing with a hint of mace in the filling.

A pinch of mace is good in buttermilk pie (an old-time Texas favorite), and in buttermilk custard or buttermilk ice cream. If you don't like the look of the dark powder in these white desserts, warm the buttermilk with a blade of mace, then set it aside to infuse until you are ready to use it; remove the mace when you begin making the dish.

This spice is not just for sweets. Curries of all kinds benefit from mace. It is standard in frankfurters and English potted meats, and a pinch of mace is most effective in shellfish dishes. Use only a small pinch—not so much that diners ask, "What's that flavor? Did you put nutmeg or something in here?" but just enough that they say, "Hmmm, this is extra good!"

Blade mace is better than ground when you want to keep a liquid clear, and it is thus traditionally used in pickling. It is also ideal for cold fruit soups or for creamed vegetable soups. A blade of mace enhances a pot of stock. Use this form of mace also for fruit or meat jellies. The blade is removed when the dish is ready.

In a pinch, nutmeg and mace can be substituted for each other, but not in equal quantities. If your recipe calls for one-half teaspoon of ground mace, use about five-eighths teaspoon (one-half teaspoon plus a pinch) freshly grated nutmeg instead.

Like nutmeg, mace can be toxic in large doses (see **nutmeg**). Note, however, that the chemical Mace spray used for personal protection and to subdue riots is totally unrelated to the spice.

mahaleb (*Prunus mahaleb*): This enchanting spice comes from a small cherry tree, sometimes called St. Lucie cherry or perfumed cherry, which grows both wild and cultivated over a wide area from the eastern end of the Mediterranean to southern Europe, as well as in parts of North America. In Europe and America, the mahaleb cherry tree is used as a rootstock in orchards, and its fine, fragrant wood is used in cabinetmaking; but Middle Easterners prefer to use it as a source of spice.

In the spring, the tree is covered with snowy white blossoms, followed by small black fruits. Each cherry contains a hard, round pit, and within each pit is an astonishingly aromatic kernel. These tan, pointed, ribbed kernels—the spice mahaleb—lend their strong, clean scent and warm, mouthwatering flavor to bread doughs, cookies, and pastries of all kinds. It is popular in Greece (under the name *malepi*), in Turkey (*mahleb*), and in Syria, Lebanon, Israel, Palestine, Jordan, and Arabia (where pronunciation varies from *mahalab* to *mahlab* to *mihlib*). Important breads, such as the Greek Easter braid, wedding pastries, or the Holy Bread of the Eastern Orthodox Church, all have their significance underscored by the presence of mahaleb.

The spice is available, either as whole kernels or as a powder, in almost any shop specializing in Greek or Middle Eastern foods. The kernels can be pulverized in an electric coffee grinder or, with a little elbow grease, with a mortar and pestle. Grind them as fine as possible, and push the powder through a sieve, discarding any remaining large pieces.

The powder is added to the dry ingredients when making baked goods. A typical recipe will call for a teaspoonful of mahaleb for approximately three cups of flour, but the amounts do vary. Rarely, a recipe will instruct you to boil up the whole kernels in water or other liquid, and then discard them, using the flavored water to make the dish.

Mahaleb's unique perfume is good with all sorts of sweets. It can even be added to white sheep's cheeses, such as feta, by stirring a little mahaleb into the brine in which the cheese is soaked. About a teaspoonful is sufficient for a pound of cheese. Armenian cooks often use mahaleb to season *basdek*, a mixture of fruit juice, sugar, and cornstarch, which is used creatively to make puddings, fruit "leathers," or a thick coating for nuts on a string—the fanciful sausagelike *roejig*.

The scented leaves of the mahaleb cherry are occasionally used as a flavoring for sauces, and for milk and other beverages.

marigold/pot marigold/calendula (*Calendula officinalis*): These attractive herbs brighten many a garden with blooms ranging in color from bright yellow, to gold, to deep orange. Although their compound inflorescences resemble single flowers, each one is actually composed of many tiny flowers, or florets, of two types: the disc florets making up the center, and the colorful ray florets that appear to be petals and are generally called that.

Whole fresh marigold flowers of any type may be used to make a stunning garnish, but more often the petals, or ray florets, are plucked off and dried. If a petal reveals a white "heel" at the base, this should be snipped out with scissors and discarded. Only the petals are eaten, and these may be used fresh or dried. The centers are tough and fibrous, with a less appealing taste. The flowers are often dried whole, and when you buy dried marigolds, it is up to you to pluck off the long, colorful ray flowers and discard the gray-brown centers. Keep dried petals in a sealed container away from heat and light. Just before using, chop the dried petals fine with a knife or, easier yet, buzz them in a small electric grinder. Fresh petals should also be chopped up, and are good sprinkled lightly over salads.

Marigold petals impart a pale yellow color to food, and are an age-old substitute for **saffron**—although their flavor is different from saffron's and much less intense. The petals also lack saffron's heady aroma, offering instead a rich musky scent of their own. They are a traditional coloring for butter and cheese, and marigold petals are sometimes fed to chickens to color their fat a rich, appetizing yellow.

Marigolds have long been esteemed for their effect on soup. A pinch or two of finely chopped petals make an excellent seasoning for any soup; this pleasant, earthy flavor will enrich any broth that seems a little too thin or watery. It is also good in pot roasts or stews.

Don't stint with marigold petals in rice dishes: A quarter cup of chopped petals for each cup of rice is generally not excessive. Chicken broth, chopped onion, or grated cheese are compatible with marigold rice. Use marigold in sweet rice pudding, too; a sprinkling of cinnamon goes well on this dish.

Mix lots of marigold petals with cream cheese, and thin with a little milk to make a dip for raw vegetable nibbles. Or chop up the vegetables and stir them into the marigold cheese (without thinning) to create a spread for crackers. A compound butter with finely chopped marigold petals makes a bright topping for cooked vegetables of all kinds; use about one tablespoon of marigold petals for four table-spoons of butter, and add a pinch of **salt** if the butter is unsalted.

Marigold may also substitute for saffron in recipes for baked goods, yielding an unusual, not-too-sweet flavor that is good in muffins, pound cake, or tea cakes. Much more marigold is required than saffron: two tablespoons of marigold petals replace one *tea-spoon* of lightly crushed saffron threads. Like saffron's, this flavor can be overdone in baked goods.

Coriander seed, **ginger**, and **coconut** are wonderfully compatible seasonings with marigold. Try adding a pinch of ground coriander seed to your marigold-butter, or serving ginger preserves with your marigold muffins, or icing a marigold cake with coconut. (Use unsweetened flakes if at all possible.)

Other flowering plants are also called marigolds and are used in cooking, the most common being members of the *Tagetes* genus, such as the French marigold, *T. patula*; the signet marigold, *T. tenuifolia*; and the Mexican mint marigold, *T. lucida*. These have composite flowers resembling the calendula bloom, and they are edible and may be used interchangeably with calendula in all the uses described above.

Note that other members of the *Tagetes* genus, while technically edible, may not taste particularly good. Test your marigolds before you use them in cooking. (Incidentally, the leaves of the Mexican mint marigold—also called Mexican marigold mint, *yerbanis* or *anisillo*—have a decided **anise** flavor and are often used as a substitute for **tarragon**.)

The extraordinary cuisine of the Republic of Georgia has embraced the marigold, sometimes calling it "Imaretian saffron" after a province of the country. The bright dried petals are a part of the national herb and spice blend, **khmeli-suneli**. In Georgia, both calendula and *Tagetes*-type marigolds are used in cooking.

marjoram/sweet marjoram/knotted marjoram (*Origanum marjorana*): Marjoram is closely related to **oregano** and the flavors of these variable plants are quite similar, but when you want a sweeter, milder taste, marjoram is called for.

The fragrant marjoram plant is very versatile. Its small velvety leaves slip nicely into a green salad. Or you might prefer to use the leaves in a vinaigrette or to make an herbal vinegar for dressing the salad (see Flavored Vinegars in the "Flavor Combinations" chapter). Marjoram perks up almost any vegetable dish. Remember this sweet herb when cooking potatoes, peas, or onions, and especially spinach and zucchini. Add the herb toward the end of the cooking time to maximize its delicate flavor. A pinch of marjoram belongs in almost any seafood dish. Clam chowder, baked fish, crab cakes, and croquettes all profit from a hint of its fine flavor. The herb is also excellent in meat dishes, and may be mixed into meatloaf and sausages, or into stuffing for chicken, turkey, and other birds. A compound butter of dried marjoram is splendid to melt over grilled chicken or fish.

Dried marjoram is as good or better than fresh, and the dried leaves hold their flavor well. Fresh leaves may also be frozen: the flavor will survive, though the looks and texture of the leaves will suffer.

Marjoram's muted flavor makes an essential contribution to the traditional French herb blend, **herbes de Provence**.

mastic/mistki (*Pistacia lentiscus*): For millennia, the bulk of the world's mastic has come from a single Greek island in the Aegean close to Turkey, known to the Greeks as Chios and to the Turks as Sakız Adası, meaning "mastic island." This aromatic resin is produced by a small evergreen tree related to the pistachio tree. When the trunk and larger branches of the tree are slashed, the mastic resin exudes from the incisions and hardens into small translucent lumps.

The delicate resinous taste of these lumps makes them an esteemed spice. All over the Middle East, mastic adds its unique taste to milk puddings, and it enriches a sweet apricot pudding made from sheets of dried apricot "leather." Mastic also provides another flavor option for the popular jellylike candy called *lokum* or Turkish delight. And in the Middle East, mastic is as common a flavor for ice cream as vanilla is in America.

In Arab countries, where it is known as *mistki*, a little mastic may be mixed with ground dates and butter—and some chopped walnuts, if you like—to create a sumptuous filling for cookies.

Mastic, often paired with **mahaleb**, is used to make sweet breads and pastries truly special. For example, it is traditional in Greek Easter bread (a seasonal specialty that is popular all year round just because it tastes so good); and in Cyprus, mastic is always incorporated into the dough for *vasilopitta*—a cake made to celebrate the New Year. And for Easter, the triangular pastries *flaounes* are baked with a sweet cheese filling flavored with mastic and dried **mint** (an inspired combination).

Mastikha, an alcoholic drink of Chios similar to ouzo, is flavored with mastic as well as **anise** seed, and a nonalcoholic drink is made by stirring a spoonful of sweet mastic jam into a glass of water.

Try dropping a single nugget of this spice into any meaty stew, soup, or broth: it will enrich the entire dish without drawing attention to itself. Many Saudi cooks practice this ancient and excellent culinary trick.

The word "mastic" is akin to our word "masticate," meaning "to chew," and you might say that this substance is the original chewing gum. Chewing mastic is believed to polish the teeth and perfume the breath. Today the Chios Gum Mastic Growers Association produces candy-coated chewing-gum rectangles that you can find in little boxes at Middle Eastern shops in America. The product is rather tough, so masticate it carefully.

Most Middle Eastern shops also sell lumps of pure mastic for culinary purposes. They are rather expensive, but you don't need very much—less than half a teaspoon suffices for almost every household recipe. Despite the fact that the word "mastic" has been used to indicate a fashionable pale yellowish-tan color, for the best flavor you should look for mastic with as light a tint as possible.

Sometimes called gum mastic, this substance is technically a resin. It will *not* dissolve in water, but it will melt to release its flavor when simmered in a hot liquid, such as broth or milk. Pulverize a few lumps in a mortar, and sprinkle a measured amount of the powder over the hot liquid. If your recipe does not call for a hot liquid, mix the ground mastic with the dry ingredients, or with chopped dates, figs, or other fruit to be made into fillings or preserves. Note that it is helpful to add a little salt, sugar, or flour, whichever your recipe permits, to the mortar before pounding the mastic, to keep it from sticking.

Some of the melted resin may stick to your cooking pot. Soaking the pot in cold water is ineffective, but you can melt this residue off by rinsing the pot several times in very hot water.

mints (*Mentha* species): The mints are extremely mutable plants, producing numerous varieties, forms, sports, and crosses. Because of their many variations, there are far more names for mints than there are actual species. The late Edgar Anderson, eminent botanist at the Missouri Botanical Garden, once remarked that it is difficult to grow mints without losing faith in botanists. If you grow mints in your garden, do not let them flower because many of the hybrids have an unpleasant flavor.

spearmint: The most important culinary mint is spearmint (*Mentha spicata*), also known as garden mint. Spearmint leaves are suitable for all the standard uses of mint, from the well-known mint jelly accompanying roast spring lamb, to fresh peas with mint, and on to mint julep and mint chutney.

Mint jelly is usually based on apple jelly with finely chopped mint leaves and green coloring added. Mint sauce, another popular condiment with lamb, is quite dissimilar; this sauce is made of fresh chopped leaves simmered with a little sugar and dilute malt vinegar (see Flavored Vinegar in the "Flavor Combinations" chapter), and salted to taste.

A pat of lemon-mint butter is excellent melting over hot baked or broiled fish; add it after the fish has been cooked, so that the tender herb leaves do not burn. (See Compound Butters in the "Flavor Combinations" chapter.)

Fresh spearmint is good with vegetables. Sprinkle a little chopped mint on cooked carrots or new potatoes in their jackets. Add a good pinch of chopped mint leaves to the water when you boil cabbage. Any legume dish—peas, beans, or lentils—is grateful for a hint of mint; add about half a teaspoon of dried mint when soaking dried beans or peas and garnish them with fresh mint leaves when the dish is served.

Mint has a well-known affinity for fruit. A sprig of fresh mint on a fruit cup invigorates the dish. **Lemon** and mint are always exciting together, in a salad dressing or in the compound butter mentioned above, or made into a sweet with sugar or, better yet, honey. You can make a mint-flavored gelatin for fruit salad or dessert by pouring a hot infusion of a generous quantity of mint leaves over plain gelatin; add a couple of drops of green food coloring if you want to emphasize the mintiness, and garnish with a sprig of leaves.

The marriage of mint and **chocolate** is famous, and justly so! Mint candies beg for a chocolate coating. Bits of mint candy liven up chocolate ice cream, and chocolate chips add depth of flavor to mint ice cream.

Mint does not participate much in classic French cuisine (although peas with mint are a standard dish), perhaps because of the tenets firmly held by classic French chefs that mint and garlic are not good together, and that mint is "antagonistic" to wine.

Italians are fond of *mentuccia*, a sweet-flavored form of spearmint that grows wild over much of Italy; the chopped leaves, fresh or dried, add an aromatic interest to breadcrumb stuffings, moistened with olive oil and lemon juice, for artichokes and other vegetables. A little chopped mint is sometimes added to minestrone to give it personality.

The Mexican *yerba buena* (or *hierba buena*) is a wild spearmint, used fresh or dried for flavor and as a garnish (but note that this common term of approbation has been applied to other herbs as well). Mint is particularly favored in Oaxaca, where it is regularly added to chicken broths and soups to give them a fresh, clean smell and prevent that musty note that often creeps into chicken stocks, especially after they have been refrigerated. This happy technique does not have to be restricted to Mexican dishes: in Asia the same effect—but a different flavor—is achieved by making chicken broth with a knob of fresh **ginger** instead of mint.

Although mint is little used in the cuisines of China, Korea, and Japan, it is popular to the south with the cooks of Vietnam, Thailand, and other parts of Southeast Asia. Various tropical forms of spearmint and water mint are used, along with an assortment of other fresh-flavored, minty herbs such as *rau ram*, called Vietnamese mint, *Polygonum odoratum*. It is also known as Vietnamese coriander. (See the **coriander** entry for more details about this sprightly herb.) In this region, the compatibility of mint with beef is celebrated, and accordingly the popular Vietnamese *pho*, a lush beef and rice-noodle soup, is generously topped with mint leaves, along with **basil**. Mint also goes into the Vietnamese fresh spring rolls. The Thais mix handfuls of shredded mint leaves (see Chopping Herbs in the "Culinary Practice" chapter) with stir-fried beef, as well as other meats, poultry, and seafood; if you are using dried mint, add it to the wok for the last minute of cooking. Sprigs of mint are a popular garnish for Thai fish dishes, and the chopped leaves are often stirred into chicken curry just before serving.

In India, mint chutneys are much loved. Variations blend mint with **ginger**, **sesame seeds**, **coconut**, mango, or tomato; of course, all of them include a certain amount of **chiles**. Mint chutney can be mixed with yogurt to make an excellent sauce for lamb kebabs. Indian recipes for baked or steamed fish often include fresh mint, **coriander**, **chives**, and other seasonings as stuffings or as toppings.

Mint holds an important place in Middle Eastern cookery, partly because of its affinity for yogurt and for white sheep's-milk cheeses such as feta. Mint also marries well with cucumber, and it is generously sprinkled over cucumber salads. These felicitous culinary partnerships are exploited in a popular summer dish combining cool yogurt with chopped cucumber and mint; a little **garlic** and a pinch of **salt** may be added. Tasty *böreks*, little pastry pockets, may be filled with a mixture of beaten egg, white cheese, and chopped mint, then deep-fried.

Tabbouleh, the popular Middle Eastern salad made with a little cracked wheat and a lot of chopped parsley, is often lifted to a higher flavor note by the addition of chopped fresh mint; the salad is dressed with lemon juice and olive oil—always a good combination with mint.

In Persia, a meal often begins (or ends) with a plate of crisp green fresh herbs, such as mint, **chives**, **dill**, **parsley**, **tarragon**, watercress, or whatever is at its best in the season. No dressing is needed. These may be eaten with bread and cheese, or simply nibbled alone at the end of the meal to aid digestion.

A sprig of fresh mint leaves makes any cool beverage more refreshing; this treatment benefits fruit punches, sangria, and iced tea, and is important in "Saudi champagne"—a tasty nonalcoholic dinner drink made from equal parts of apple juice and sparkling water poured over lots of lemon slices and topped with mint leaves. Lots of mint is also used for the culturally significant Moroccan mint-flavored green tea, which is flavored and sweetened in the pot, and then poured dramatically and aromatically from a height into small glasses whenever Moroccans want to relax, celebrate, or extend hospitality.

As for the famous mint julep, the more mint the better. Use at least four sprigs of fresh mint for each tall glass or cup. (Silver Julep cups have been incorporated into the mystique of the drink.) To bruise or not to bruise the mint leaves: that is the question you must answer for yourself. But either way, put the mint in the bottom of the cup with about a teaspoon of superfine or confectioner's sugar, add a little water until the sugar has dissolved, then pack the cup with crushed ice. Chill the cup or glass until it is completely frosted, and finally pour in three to four ounces of bourbon. Garnish the top with even more sprigs of mint before serving.

Be adventurous with mint! For variety to spice up your dinners, try replacing basil with mint in pesto, or cilantro with mint in salsa. Put a few fresh leaves, or a big pinch of dried leaves, in tomato sauce. Make finger sandwiches of chopped fresh mint and cream cheese for an afternoon tea, or incorporate finely minced mint into orange marmalade at an elegant brunch. Use a little fresh mint in potato salad and coleslaw. Tear up a few mint leaves and toss them into your green salad, and use sprigs of mint instead of curly parsley as a garnish. If the mint flavor is good but seems too strong, you can tone it down by mixing in a little parsley.

For a mild form of mint, you might choose the round-leafed apple mint, a nonvariegated form of *Mentha suaveolens*. Another mild, fruity mint is the broad-leafed orange mint, formerly *Mentha aquatica citrata*, also called bergamot mint (see **monarda**). The leaves of orange mint are particularly good in fruit dishes: try sprinkling them over orange slices on a bed of lettuce for a salad. If you have any other type of mint whose flavor pleases you—be it called basil mint, pineapple mint, silver mint, or whatever—by all means use it to your taste buds' delight; but please do see the paragraphs below on peppermint.

dried mint: Mint dries very well (see Drying Herbs in the "Culinary Practice" chapter) without becoming musty. The taste and effect of dried mint is rather different from that of fresh mint, however, with dried mint providing a backup flavor note and fresh mint taking the lead. Some recipes even call for both fresh and dried (see, for example, the legume dishes described earlier in this section and the Lebanese *fattoush,* which follows). Almost all dried mint commercially available is spearmint.

Dried mint is even more popular in Middle Eastern cuisine than the fresh herb. It often seasons fabulous meat dishes, such as ground lamb fillings for cabbage leaves, or meat-filled phyllo pies, or meatballs cooked in a stew, or a North African tajine or Moroccan mutton soup. Not only lamb or beef benefit from mint; it also improves braised chicken and vegetables.

Dried mint also enhances the popular vegetarian rice stuffing with chopped onion, currants, and pine nuts that Middle Eastern cooks use to stuff whole tomatoes, bell peppers, zucchini, and other vegetables, as well as grape leaves. Dried mint is added to the dressing made of lemon juice and olive oil for the Lebanese bread salad called *fattoush,* frugally assembled from pieces of stale or toasted pita bread, cucumbers, tomatoes, green onions, lettuce, and other greens, including fresh mint leaves as well; garlic may be added if desired.

Dried mint flavors the refreshing cold drink of Iran, *doogh,* made by thinning yogurt with soda water. Persians also like to garnish their thick winter soups or stews with a design of "burned mint," that is, powdered dried mint which has been quickly fried in butter or oil; it is cooked just until the color is dark but the taste remains sweet.

Turkish cooking uses a lot of dried mint, sprinkling it as a garnish on yogurt dishes as well as cooking it into pilafs, lentil dishes, and yogurt soup and incorporating it into a tasty *kısır*—a cold side dish of cracked wheat reddened with ground chiles and sweet red peppers, seasoned with dried mint, finely chopped parsley, olive oil, and lemon juice. The ancient Turkish dish *mantı*—tiny meat-filled "ravioli" in a yogurt sauce—is often sprinkled at the table with a topping of dried mint or, better yet, with a bright mixture of dried mint, black **pepper**, and **sumac**.

peppermint (*Mentha* x *piperita*): Peppermint is a natural hybrid of spearmint and water mint (*Mentha aquatica*). This special form of mint cannot be grown from seed, but must be rooted from cuttings. While spearmint is an ancient culinary friend, a subject of Greek myth and mentioned in the Bible, the hybrid peppermint has only been noticed and encouraged for the past three centuries. In that short time, however, peppermint has become one of the most popular flavors in the world. Peppermint is cultivated primarily for its pungent essential oil, which is used extensively in medicines, mouthwashes, toothpastes, and chewing gum—as well as in peppermint confections. It is not used in cooking savory dishes.

Peppermint tea, prepared by steeping the leaves in hot water, is a refreshing drink that may be served hot or iced. It has a reputation for soothing stomach upsets, but because of the high menthol content of peppermint oil, this drink should not be given to infants. (Spearmint oil is almost entirely lacking in menthol.)

To make peppermint confections, you can buy oil of peppermint at specialty food shops or from your druggist. Be sure to tell the druggist that you want a culinary grade oil of peppermint, suitable for flavoring, and not something intended only for potpourri and aromatherapy. Do not buy Japanese peppermint oil, even though it is cheaper, because its flavor is inferior; this mint (*Mentha arvensis*) is grown for its very high menthol content.

Oil of peppermint is extremely strong. Many recipes call for only a couple of drops of peppermint oil, sometimes only one drop. Measuring drops is difficult without an eyedropper. Never attempt to pour out the oil drop by drop directly into the pot or bowl; always collect the proper amount of oil in a spoon first and then transfer it to the other ingredients. A useful guide for measuring is to keep in mind that about eight to ten drops make one-fourth teaspoon. If you spill any oil of peppermint, wipe it up immediately; wash it off your hands well before touching your face and be careful of your eyes. Do not put this potent oil on your tongue to taste it; taste a bit of the entire recipe instead.

Peppermint extract is made by mixing oil of peppermint with alcohol and water. The high alcohol content makes the extract flammable, so it should not be poured out near your gas burner or other open flame. Peppermint extract is much milder than peppermint oil and better suited to baking. Try mixing two teaspoons of pepper-mint extract into a cake batter; this is especially good in chocolate cake. The extract is also a good choice for making ice cream.

An easy and attractive way to incorporate peppermint flavor into frostings, ice creams, or cookies is to crush peppermint candy into small pieces. The result is pretty as well as tasty.

When making peppermint-flavored sweets, you will probably want to add a few drops of food coloring, both for looks and as a signal to the eater. Although the standard color given to spearmint and other mints is green, peppermint is traditionally associated with a bright pink or red.

pennyroyal *(Mentha pulegium)*: Pennyroyal is really too strong—both in taste and its physiological effects—for culinary use, but it befriends the cook in the herb bed. A few plants growing here and there in the garden will keep ants and other insects away from a place where you dare not risk using toxins.

monarda/bergamot/bee balm/Oswego tea *(Monarda didyma)*: A native American, this tall herb has splashy flowers in shades of red and pointed, oval leaves with a red center vein. The entire herb is strongly scented with an aroma resembling that of the small, sour bergamot **orange**. It has long been used to make a refreshing tea with reputed medicinal properties. Both leaves and flowers may be brewed up fresh or dried. To make an infusion, use about half a dozen of the leaves, or the long flowers from two or three flower heads, depending on size, for each cup of boiling water; one and one-half teaspoons of dried leaves and/or flowers per cup yields about the same result. (See Teas and Tisanes in the "Culinary Prac-tice" chapter.)

These attractive flowers and leaves make nice garnishes, and are sometimes finely chopped and added sparingly to fruit or green salads to good effect.

In 1996, this herb was rescued from obscurity when the Interna-tional Herb Association declared monarda the official herb for Na-tional Herb Week. Suddenly it was a culinary star, featured in many

imaginative, delectable new dishes. Its citrus flavor was found to be excellent with duck, chicken, or turkey. Finely chopped fresh leaves can be added to stuffings for these birds or incorporated into gravy and sauces. Monarda also brightens pork dishes and is a welcome seasoning for homemade sausages.

Monarda flower petals or leaves bring an interesting flavor to cookies, meringues, and breads. Fruit punches are enhanced by the herb. Or make an infusion of monarda with hot milk, sweetened if desired, for a delightfully different bedtime drink.

mustard: There are three basic types of mustard seeds—black, *Brassica nigra*; brown (or Indian), *Brassica juncea*; and yellow, *Sinapsis alba*. This terminology refers to the color of the outer hull; the insides of all three of them are pale yellow. Black seeds are by far the hottest and yellow seeds are the mildest. Sources that call for "white mustard seeds" mean the yellow ones, and modern references to "black mustard seeds" probably refer to the brown variety, as the true black seeds have become increasingly hard to find in the past half-century, due to the fact that brown mustard seeds have proved to be more suitable for mechanical harvesting.

Whole yellow mustard seeds are an important component of **pickling spice** because of their excellent preservative qualities as well as their flavor.

Whole seeds are also used liberally to season Indian vegetable dishes. Indian brown mustard seeds are a variety with a different flavor than the brown seeds used in Dijon and other European mustards. The seeds should be lightly fried in ghee or oil, or else toasted in a dry pan before they are added to the dish. (See Toasting Spices in the "Culinary Practice" chapter.) Be sure to keep a lid on the pan or the seeds will pop out when they get hot. This heating makes the seeds quite mild, with a nutty flavor.

In the eastern corner of India—Bengal and neighboring regions—mustard is almost a staple. It lends its distinctive flavor to many Bengali dishes, especially fish. Many vegetable and fish dishes call for crushed mustard seeds in their preparation, then are cooked in hot mustard oil at the end. Freshwater fish, smeared with a paste made of brown mustard seeds, **chiles**, and **turmeric**, are deep-fried in smoking mustard oil for a prized dish. Remember that mustard seed and mustard oil become milder when heated. If you are making a fish curry, take a tip from the Bengalis and lightly sizzle your **curry powder** in a little mustard oil before adding it to the pot.

powdered mustard: Powdered mustard seed, also called mustard flour or ground mustard, has no aroma and is virtually tasteless when dry. Its flavor and pungency develop after it is mixed with liquids, as you can experience for yourself if you let a little dry powder sit on your tongue for a short while. Strangely enough, the addition of a small amount of powdered mustard serves as a remarkable flavor enhancer in both sweet and savory foods. Perk up boiled root vegetables such as carrots, potatoes, parsnips, and **parsley** roots by adding a half teaspoon of mustard power along with the salt in the cooking water. One-half teaspoon of powdered mustard included with the dry ingredients in the batter for **chocolate** cakes, brownies, or cookies will intensify their chocolate flavor but will be otherwise undetectable. Similarly, a little dry mustard in the fruit filling for a pie or a cobbler will enhance its fruitiness. To bring out the best in coffee, try the old French trick of putting a tiny pinch of mustard powder in the bottom of a cup before pouring in the hot liquid.

prepared mustards: Americans eat more mustard than any other spice but black **pepper**, and they have a vast assortment of mustard concoctions to choose from. Prepared mustards are basically a paste of water, vinegar, and ground mustard seeds. "Country-style" mustards contain coarsely ground seeds. Vinegar, or some other acid such as wine, is necessary if the prepared mustard is going to be kept longer than a few hours; otherwise the flavor of the mustard will dissipate. If you make your own prepared mustard for immediate use, mix the mustard powder with cold water and allow it to stand for a quarter hour for the flavor and pungency to develop. Similarly, if you are using whole mustard seeds, soaking them first in cold water will give a much better flavor. Prepared mustards can be varied endlessly by the type and quantity of mustard seeds, by the choice of vinegar (see Flavored Vinegars in the "Flavor Combinations" chapter), and by the addition of herbs—**tarragon** is a favorite—or spices such as green peppercorns, or distinctly flavored sweeteners such as maple syrup, caramelized sugar, or honey.

American hot-dog mustard is made from yellow mustard seed, but its bright color comes more from **turmeric** than from the mustard. English prepared mustards are smooth yellow pastes like the usual American mustard, but English mustard is hotter, being made from both brown and yellow mustard seeds. In England, eating roast beef without mustard is almost unthinkable.

Oriental mustard, popular in Chinese and Japanese cuisine, is also smooth and bright yellow. It is made very hot, usually entirely from brown seeds, to serve as a dip for fried egg rolls and other foods. Traditionally, this mustard is prepared for immediate consumption by mixing powdered brown seeds with water and allowing the mixture to stand for a few minutes to develop its pungency. Occasionally, flat beer, sake, or rice wine vinegar may be used instead of water. With little or no acid, this mustard will not hold its flavor and sharpness, and it should be prepared in small quantities.

There are several types of French mustard, generally identified by their places of origin: Dijon, Orléans, Bordeaux, Meaux, and so on; these are all made with brown seeds, ground to different consistencies, and moistened with white wine, wine vinegar, or other grape products such as verjuice or must. In menu French, meat and poultry dishes labeled *à la diable* ("deviled") have a pungent seasoning that usually includes prepared mustard.

German mustards are smooth, mild, and sweet—perfect to slather on sausages.

A little prepared mustard is often incorporated into mayonnaise, both as a seasoning and as an emulsifier. A smooth prepared mustard, with a pungency that suits your taste, will complement cheeses, cheese sauces, and Welsh rabbit. Country-style mustard is surprisingly good for basting grilled or roasted pork and other meats; you can be lavish with it because the heat reduces its pungency, leaving only a pleasing savory flavor. To serve with a mild food such as fish, prepared mustard can be toned down with the addition of a little cream.

Mustard sauces and dressings are made by adding prepared mustard to a standard base or emulsion. Honey and mustard together make a pleasing spicy-sweet combination; taste to see that the flavors are balanced.

mustard oil: Mustard seeds are about one-third oil. Raw mustard oil is hot and peppery, but when heated, it gradually turns pale in color and mellow in flavor; you can use this fact to adjust its pungency to suit your dish and your diners. If you wish to eliminate its sharpness, heat the oil to the smoking point for a few seconds; thereafter, it remains mild, whether used hot or cold. Properly adjusted, mustard oil lends welcome flavor as a dressing both for cold salads and for hot vegetables such as green beans.

Mustard oil is an excellent cooking oil. Not only is it a favorite of the Bengalis, but it is often used in Kashmir as well. Try using it to sauté vegetables for a fine flavor variation.

Perhaps the most unusual mustard dish is the Italian *mostarda di frutta*; this is sometimes called Cremona fruit, although a few other cities also have their versions. This tasty, chutney-like concoction is made from a selection of a wide variety of ripe fruits—apricots, pears, cherries, melon, figs, plums, and so on—which are candied in sugar syrup or reduced grape must, then flavored with mustard oil; **garlic** and other spices may be added. "Mustard fruit" is served as a relish with cold meats, game, goose, and other strongly flavored foods such as eel. If you like honey-mustard dressing, you'll love mustard fruit! In the absence of this specialty, a similar effect, although a different flavor, can be achieved by substituting candied **ginger** and its syrup for the candied fruit.

Mustard oil *must* be stored in the refrigerator to avoid rancidity.

mustard greens: Cultivars of the brown mustard plant, *Brassica juncea*, generally supply the strong-flavored leaves eaten as a vegetable, although it is seedlings of yellow mustard, *Sinapsis alba*, that participate in the traditional English salad and garnish called "mustard and cress."

Mustard greens are edible cooked or raw, but only the tender, very young leaves are mild enough to be used in a salad. The greens should not be cooked in an aluminum or iron pot.

N **nigella seed/onion seed/kalonji** (*Nigella sativa*): This unfortunate herb has been saddled with a number of inaccurate and misleading names: black sesame, black cumin or wild cumin, Russian caraway or black caraway, and even fennel flower and nutmeg flower. Its tiny, angular black seeds are often called onion seeds, although the plant is unrelated to onions; this is because nigella seeds look like onion seeds, not because of any oniony flavor. Indian recipes call the seeds *kalonji*, and in Russia and parts of Eastern Europe they are *charnushka* (or *chornushka*), meaning "the little black ones." Note that the Latin botanical name of the plant also refers to "little black ones." This succinct scientific name offers a straightforward way out of our nominal confusion: let's just call the plant "nigella" and its useful product "nigella seeds."

Nigella seeds have a sprightly flavor that is both sour and pleasantly bitter. They are an important spice in Turkey, India, and the broad area in between, extending up into Russia and down into Arabia; in this area, they are traditionally sprinkled on breads and savory pastries of all kinds. The seeds sometimes top the treasured sweet *aşure*, or Noah's pudding, a kind of wheat-berry pudding with dried fruits and nuts, said to have been put together by Noah's wife after the Flood from all the leftovers on the Ark. A black nigella seed here and there is tasty and attractive when woven into white Armenian string cheese. From the Russian tradition, this spice has traveled atop Jewish rye bread to Israel and to New York City.

The flavor of nigella seed is not strong, but it is unique and pervasive, so only a few widely scattered seeds are sufficient; using too many yields an unpleasantly acrid odor and taste.

Nigella seeds often need a bit of cleaning before you use them. Spread out one or two spoonfuls of seeds in a pie plate or flat-bottomed bowl, and pick out any light-colored bits of stem or other plant parts. Then gently trickle a little cold water into the dish to rinse the seeds. Pour the seeds and water through a strainer and spread the spice to dry on a clean cloth or paper towel. Do not put any washed seeds back into the spice jar! This falsely frugal practice invites mold into the entire jarful. It is safer to discard any unused washed seeds.

To make the seeds adhere to loaves of bread, first brush the top of the unbaked dough with beaten egg yolk, then scatter on the seeds. If you want to mix the seeds into the dough instead, crush them slightly beforehand, and use only about one-fourth teaspoon of seeds per cup of flour.

In India, nigella seeds often invigorate bland foods such as lentils, beans, and some vegetables and pickles. Toast the seeds lightly before adding them to these dishes (see Toasting Spices in the "Culinary Practice" chapter). Nigella seeds are one of five standard spices in the Bengali mixture **panch phoran**.

Despite reports that nigella seeds have been used as a substitute for black **pepper**, their flavor is actually more penetrating and persistent than peppery, and such a substitution is not successful.

nutmeg (*Myristica fragrans*): The tall, lush evergreen nutmeg tree has shiny leaves and a pretty pale yellow fruit that resembles a peach in appearance but not in texture, as its flesh is fibrous and husky. As this fruit ripens on the tree, it splits open along a vertical groove, revealing the dark shell of the seed covered by a scarlet net, which is the spice **mace**. The kernel of the seed is a whole nutmeg. With the mace removed, the nut is dried until the nutmeg rattles around inside its shell. Then the shell is broken and discarded, leaving the light brown, woody nutmeg, which varies in shape from a long oval to round.

The nutmeg tree grows primarily in Indonesia, where it is native. Malaysia, India, and Sri Lanka also produce a certain amount of this spice, and a large number of trees are grown on Grenada and other islands of the West Indies. Nutmegs from the Eastern and Western Hemispheres differ in oil content and thus in flavor. Western nutmegs have a very high fixed-oil content, so high that they are unsuitable for commercial grinding. On the other hand, they have less of the volatile essential oils than the eastern nutmegs, and are therefore milder in flavor. Which nutmeg is better is strictly a matter of taste, and spice dealers proclaim their opposing choices with equal pride.

Neither nutmeg nor mace is much used in the cuisines of the countries that produce them, although they are esteemed as medicine. The Caribbean islands do use some nutmeg as a seasoning for jerked meats, and in curry powder. In some of the regions where nutmeg is grown, the local people cook the fibrous fruit in sugar syrup to make a sort of candy or jelly, which tastes pleasantly of nutmeg.

Nutmegs gathered in the wild from other, related species of trees occasionally come on the market with names such as Macassar, Papua and Bombay nutmeg; but these are feeble and pale, lacking the fragrance and flavor of the real thing.

Although ground nutmeg is available from the supermarket, it is preferable to purchase whole nutmegs. These soft, mottled seeds are easily grated directly into the dish as you prepare it, giving you full measure of the warm, heady aroma and lingering tongue-tingling taste. Nutmeg graters come in all shapes and sizes, but any rough, raspy surface is sufficient to scrape off a sprinkling. In eighteenth-century England and America, gentlemen carried elegant silver nutmeg graters on their watch chains to ensure a ready supply of nutmeg for topping off their ales, possets, and punches.

Whole nutmegs can also be cut into quarters, one or two of which can be boiled with cream, sugar syrup, broth, or some other element of a recipe to impart a good nutmeg flavor.

Nutmeg is a good mate for vegetables, as the Dutch and Italians know. Vegetables particularly suitable for this spice are cauliflower, potatoes, spinach, white cabbage, and all kinds of squashes—which is why nutmeg just has to be in pumpkin pie. Nutmeg is good with cheese, and should be standard in macaroni and cheese along with fondue. Of course, custards and eggnog look nude without a dusting of this spice. Add a pinch of nutmeg any time you put walnuts in a dish. Apple, carrot, and zucchini cakes all yearn for nutmeg—as indeed does any kind of spice cake.

Nutmeg is nice with other spices, and it appears in several spice blends such as **curry powders**, **garam masala**, **jerk seasonings**, **mixed spices**, **mulling spices**, **quatre épices**, and **ras el hanout** (see the "Flavor Combinations" chapter). Surprisingly effective is nutmeg combined with **chiles** (as in, say, a Southwestern cornbread) or nutmeg with **thyme** (try this remarkable combination in a chicken broth; garnish with **lemon**). Consider this versatile spice also whenever you are cooking with **rose** petals, lemon, or onions.

Mace can be used in place of nutmeg, if necessary, but use a pinch less mace; that is, for one-half teaspoon of grated nutmeg, measure out three-eighths teaspoon of ground mace.

Although nutmeg is perfectly safe in the quantities called for in recipes, larger amounts should not be eaten. Consuming a whole nutmeg (equivalent to approximately four teaspoons of grated nut-meg) can lead to disorientation or hallucinations, followed by a severe hangover. Two whole nutmegs might bring on convulsions, and three could be fatal.

*O*range/sweet orange (*Citrus sinensis*): Be sure to see the Citrus section of the "Culinary Practice" chapter. The sweet, wistful flavor of orange is a favorite in candies, baked goods, sherbet, and gelatin desserts. It also melds extremely well with **chocolate**. Savory dishes also profit from orange flavor: it is a fine partner for carrots, for example. Orange flavor may be added by using the juice (or an orange liqueur), the grated zest or pieces of candied peel, or a commercially prepared orange extract.

Among nuts, walnuts make the very best combination with orange flavor, as in orange cake with walnuts, or biscotti studded with walnuts and orange peel.

sour orange/Seville orange/bitter orange (*Citrus aurantium*): Intensely orangey, the sour orange has an extremely acidic juice and a peel rich in essential oil. This is the species of orange that made the reputation of orange marmalade and orange liqueur.

One variety, the yellowish bergamot orange (*Citrus aurantium* var. *bergamia*), is grown in Sicily and southern Italy primarily for its peel, which yields oil of bergamot, an essential ingredient of Earl Grey tea and other foods, and also of fragrances.

orange flower water/orange blossom water: This exquisite flavoring is really a food perfume. Because the aroma of the food we eat is at least as important as what the taste buds detect (see the Flavor section in the Introduction), this perfume can work wonders with the flavor of a dish. Many countries around the Mediterranean produce this fragrant water, usually from the blossoms of the berga-mot or other sour orange trees. France and Lebanon produce it on a commercial scale. Of course, there are many different grades and types. Select one that pleases you—it should be as joyful as a wedding in June—and add it with a light hand after the dish has been cooked.

Orange flower water is used in Provençal cooking, particularly for desserts but also for vegetable dishes such as spinach. It can be the principal flavoring in ice cream or provide a more subtle accent in the traditional Niçoise *tourte de blettes*—a tart filled with Swiss chard, apples, raisins, and pine nuts, scented with a teaspoonful of orange flower water.

Orange flower water is indispensable in Middle Eastern cooking. It can be used anywhere that **rose** water is called for, according to preference or to vary the flavors on the menu.

Even a mundane serving of canned fruit can be lifted out of the ordinary by the addition of half a teaspoon of orange flower water.

oregano/wild marjoram (*Origanum vulgare* and other species): This Old World herb became popular in the United States after World War II along with pizza; oregano was often described as "the pizza herb." Its vigorous flavor is just right for the traditional tomato sauce. Cook the sauce with oregano, and/or rub a little of the dried leaves between your palms to sprinkle them over your slice on the plate; but don't sprinkle the herb on the pizza before baking, as the leaves will scorch.

Oregano is an excellent seasoning for other Italian dishes and for anything made with tomatoes or eggplant. Try it in recipes with mushrooms, beans, eggs, cheese, and fish. Assertive oregano is an ideal partner for olives, and many types of herbed olives take advantage of this affinity. For example, the muffaletta sandwich—a specialty of New Orleans—spreads a distinctive olive relish on a stack of salami, ham, and cheese. For the olive salad, a cup of chopped olives is seasoned with about a teaspoon of oregano, along with chopped pimento and carrot, oil, and whatever else suits your fancy.

This spice is almost always used in dried form, since many cooks feel that the fresh leaves possess too much of the camphorous note typical of oregano and **marjoram** flavors.

Oregano makes a refreshing tisane (see Teas and Tisanes in the "Culinary Practice" chapter). A hot oregano tea is good drunk alone or, for a tasty, healthy snack, accompanying herbed crackers or home-baked butter cookies made with a few toasted seeds in the dough. (Choose your favorite of **anise, caraway, dill,** or **fennel** seeds.)

Just as the essential oil containing anethole gives a so-called "**licorice**" flavor to many different herbs, so the flavor of oregano is widely distributed in the plant kingdom. Many different plants are sold as "oregano," and you should carefully taste the one you buy or grow. Some oreganos smell and taste strongly of **thyme**: be sure that this is the flavor you want before using them.

Oreganos come in different strengths also, ranging from the sweet gentleness of **marjoram** to the assertive Italian or Mediterranean oregano. When "Greek oregano" or "Mexican oregano" is called for, the recipe wants an intensely flavored herb. Mexican oregano—usually *Lippia graveolens*—is a particularly strong-flavored, sharp herb that is ideally suited to the popular Mexican *posole*, a spicy, meaty soup with hominy. Mexican oregano is the perfect ingredient in **chili powder**.

If you don't like **coriander** leaf (cilantro), or you can't find **epazote**, substitute oregano.

*P***paprika** (*Capsicum annuum*): Paprika is derived from a variety of the same species of plant that gives us sweet bell peppers and most of the hot **chiles**. Generations of Hungarians worked patiently to breed a unique, bright red, tapering pod with a mild and sweet flavor. The ripe pods are dried in the sun, then ground to a fine powder. On the dining table in Hungary, paprika always has a place alongside the **salt** and **pepper**, and sometimes it replaces the black pepper altogether. Today the United States, Spain, Morocco, and other countries produce much of the world's paprika, but this seasoning remains the soul of Hungarian cuisine. No goulash is authentic without this spice. Classic French dishes served *à la hongroise* are always made with paprika.

For variety, there are different types of sweet paprikas, labeled sweet, mild, and "delicatess" (very mild) and also a range of sharp, or rose, paprikas, which are hotter than the others but by no means incendiary; they may measure up to about 3 or 4 on a heat scale of

1 to 10 (see the discussion in the section on **Chiles**). Always assume that your recipe wants sweet paprika, unless it specifically calls for another type.

Paprika goes very well with pork, beef, veal, chicken, lobster, oysters, shrimp, scallops, and the more delicate fish, and also with mushrooms of all kinds. A dusting of paprika livens up the look and taste of cauliflower, hard-cooked or deviled eggs, cucumbers, sour cream, cream sauces, cheese sauces, and Welsh rabbit.

Lending color as well as taste and aroma to a dish, paprika is not a spice to be shy with. It is generally measured out by the heaping teaspoonful. Only the hotter rose paprika is added by the pinch, often at the end of cooking a dish that already incorporates spoonfuls of regular paprika.

Be careful not to burn paprika. Fats and oils can get too hot for it and spoil its flavor. Always add paprika to a dish after all sautéing or browning has taken place. This means that paprika should not be added to the flour in which meat will be dredged before frying. Cool a sizzling hot dish down by removing it from the heat; wait a minute or so, then carefully add a little water or other liquid to guarantee that the temperature does not exceed the boiling point before stirring in the paprika. Paprika can boil in a liquid and swap flavors with the other ingredients for as long as you like.

Store your paprika, well-sealed, in the refrigerator to retain its bright red color; otherwise, paprika tends to turn brown as it ages. Keeping it in the refrigerator also protects it from weevils, who like paprika as much as we do.

Ground red pepper or cayenne pepper does not make a good substitute for paprika, unless it is just sprinkled on top for looks. Not only is the red pepper much too hot, but its flavor is not the same. You can achieve a better approximation of the paprika flavor and effect if you temper the red pepper with a bit of tomato paste or puree, but take into account the salt and other seasonings that most manufacturers put into their tomato products.

parsley (*Petroselinum crispum*): Today most supermarkets offer two varieties of parsley—curly, and the flat-leafed variety also known as Italian parsley. These two herbs have different flavors;

most cooks agree that flat-leafed parsley is sweeter and curly parsley is milder. The two leaf types also lend very different textures to the final result, so it is a good idea to get the right type of parsley for every dish. If your recipe does not specify which one to use, consider the source: a cookbook from an English-speaking country will probably intend curly parsley, while Mediterranean and Middle Eastern sources are likely to be thinking of the flat-leafed variety. France is divided on the choice of parsley (as are cookbook authors on which of these herbs should be called French parsley); generally, in the northern part of the country, including Paris with its classic French cuisine, curly parsley is preferred, while in Provence and other southern regions the flat-leafed is more popular.

When buying parsley, watch out for cilantro, the leaf of the **coriander** plant, also called Chinese parsley. It looks a lot like flat-leafed parsley, but its taste is shockingly different, as anyone can tell you who has ever bitten into one while expecting the other. To tell the difference, brush the leaves with your hand and inhale: parsley has a mild, clean, and green aroma, while the smell of cilantro is strong and unmistakable.

To chop parsley, take a few moments to arrange a small bouquet of sprigs with all the leaves at approximately the same level. The stems will stick out at different lengths. Holding your bouquet by the stems, lay the leaves on a cutting board and slice off the leaves with diagonal cuts. (Don't discard the stems in your hand: these can be frozen for future use in making stocks.) The resulting heap of parsley leaves on your cutting board will contain only a few of the thinner stems, which may be removed or not, as you prefer. Gather the leaves together and chop them with a large knife, as fine as desired.

curly parsley: Curly parsley has traditionally functioned as a flavoring, having been for centuries one of the most commonly used herbs in the Western world. This popularity is well deserved. Parsley's fresh, green flavor is extremely agreeable, and it lifts and brightens any dish. It blends well with any other herb and serves as a moderator for strong flavors.

You can add that fine flavor to roast beef, and lighten the dish, by "larding" the roast with parsley; that is, pierce the roast all over with the tip of a sharp knife, then stuff the slits with parsley leaves, and roast as usual. Garnish the serving platter with parsley sprigs.

Parsley is also essential in the classic French combinations **bouquet garni**, **fines herbes**, and **persillade** (see the "Flavor Combinations" chapter) and in the popular *maître d'hôtel* butter (see Compound Butters in the "Flavor Combinations" chapter). Curly parsley also makes a beautiful, if too-familiar, garnish. If a garlicky meal is served with a parsley garnish, you might want to finish by eating the sprig of herb; parsley really does mitigate the smell of garlic on the breath, its only limitation being that the garlic odor is longer-lasting than its parsley antidote. If the cook's fingers have absorbed a reek of garlic or onion, the smell can be removed by rubbing the fingers vigorously with fresh parsley leaves.

Crisp fried parsley also makes an attractive and tasty garnish. Thoroughly dried sprigs are dipped in deep, hot oil for less than a minute (watch out for splattering hot oil from the moisture of the leaves), then drained well and lightly salted. Dehydrated parsley flakes in this country are made from curly parsley.

flat-leafed parsley/Italian parsley: In the Mediterranean regions, flat-leafed parsley has been used almost like a vegetable, and in a Middle Eastern tabbouleh or an Italian parsley pesto this herb is the main ingredient. Flat-leafed parsley is appropriate for the Italian topping **gremolata** (see the "Flavor Combinations" chapter). Persian cooking frequently calls for *sebze gormeh*, a mixture of many dried herbs, with this type of parsley foremost among them. Parsley is also included on the plate of plain green herbs sometimes passed around as an appetizer or a refreshing finish to a Persian meal. This form of parsley is not as good for frying as the curly variety.

parsley roots: There is a variety called Hamburg parsley that is grown for its roots. These look like little parsnips and resemble them somewhat in taste, although their flavor is brighter—less sweet and less starchy—than that of parsnips. Parsley roots are most popular in northern Europe, but they are occasionally available in America. The leaves of this type of parsley are flat; they can be eaten like any other variety of parsley, but the flavor is not as fine. Parsley roots, however, are quite appealing; they can be included in pot roasts and stews along with other root vegetables, but they are good enough to serve on their own. Parboil and candy them with brown sugar, butter, and freshly grated **nutmeg**; or stew them in a little light roux of cooked flour and butter that has been spiked with a good white wine vinegar. Both these dishes should be topped with a sprinkling of chopped parsley leaves.

pepper/black pepper (*Piper nigrum*): Plain old black pepper—that familiar and essential, pungent, aromatic taste heightener and brow warmer that sits beside the salt on nearly every dinner table in America—grows on a vine native to southern India, which proudly bears the straightforward Latin name *Piper nigrum*, that is, "black pepper." The small berries are picked just before they ripen, then dried in the sun until black and shriveled. In this form, they make excellent, airtight little containers for that characteristic biting taste and the unmistakable mouthwatering pepper aroma which is released when they are ground. After grinding, the bite remains but the aroma fades gradually away as the ground spice ages, and for this reason, it is preferable to use freshly ground pepper whenever possible. Store ground pepper in an airtight container away from light and heat to preserve its volatile aroma as long as possible.

The word "peppercorn" preserves the meaning of "corn" as a grain or other small, hard particle. For many centuries, the term has indicated cereal grains, and when English-speakers reached America they applied it to the dominant grain of the New World: maize. Also, corned beef is preserved by grains, or "corns," of salt.

Much of the black pepper sold in America consists of small peppercorns grown in Brazil. A variety from South India, Telicherry extra bold (that is, large), is in demand among the makers of salami and other hard sausages. A slice of these cold cuts reveals a beautiful big cross-section of the peppercorn with its pale starchy core in the center.

Cracked peppercorns are very effective in some dishes, providing the eater with an occasional burst of flavor and heat. You can crack the peppercorns yourself on a flat surface by rolling a heavy skillet over them, applying enough pressure to make them crack. A coating of cracked black peppercorns pressed into beef steaks or on tuna steaks brushed with olive oil makes them exciting to eat.

Ground black pepper is great in sweets, such as the German Christmas treats *Pfeffernüsse* and the plump, chocolatey Mexican wedding cookies. Sprinkle a little freshly ground black pepper over **chocolate** ice cream for a sassy flavor. And steep a few cracked peppercorns in pancake syrup to pour over a hearty breakfast of flapjacks and sausages.

The vivid aroma of pepper animates the flavor of fruits. The Gascons in southwest France cook figs and pears with it; in India it is freshly ground over slices of juicy **orange**s, and in the Caribbean, papaya is traditionally eaten with a seasoning of **lime** juice and black pepper.

Pepper and **cinnamon** complement each other beautifully. In the famous kitchens of the Ottoman sultans, **salt**, pepper, and cinnamon were a standard seasoning trio for all meat and poultry dishes. And the tradition lingers in many contemporary recipes from the Middle East.

white pepper: White pepper comes from the very same plant as black pepper, but is the result of different processing. The berries are picked at a later stage, when they are riper; they are then soaked in water until the dark outer skins can be rubbed off. When dried, they are white peppercorns, slightly milder and with a somewhat different flavor than the black ones. In most European households, white pepper is favored over black, while in America we generally use black, except in a few dishes where a lighter color might be preferred, such as in cream sauces or cheese dishes.

mignonette pepper/shot pepper: A mixture of cracked or coarsely crushed black and white peppercorns is popular in classic French cuisine and in French Canadian cookery for meats of all kinds. It can be used in cooking or set on the table at the discretion of the diner. The term mignonette is used rather loosely: sometimes **coriander** seed, **allspice** berries, or **clove** may be included in the mignonette blend, while at other times the word refers simply to cracked white peppercorns. As a derivative of the adjective *mignon,* meaning "cute" or "sweet," the name mignonette is also applied to elaborate specialties made with choice bits of meat or poultry, and to other dear little dishes.

When a mixture of black and white peppercorns is finely ground, the resulting powder is called gray pepper.

green peppercorns: Green peppercorns are, again, from the same vine as black and white pepper, but the berries are picked before they are ripe. Unlike black and white peppercorns, they are not sun dried and so do not travel well, tending to darken with age and become moldy. To reach the distant eater, green peppercorns must be pickled in brine and canned or—in recent years—freeze dried. They have piquant, fresh flavor and a bright aroma. There are many dishes where you might want to show off this wonderful spice, from meat slathered with green peppercorn butter to a chilled cucumber and yogurt salad spiked with green peppercorns. If you are using peppercorns that have been preserved in brine, rinse them off before using, then crush them in a mortar or with a rolling pin. They are soft and crush easily. If you have freeze-dried peppercorns, crush them and mix them with the butter, yogurt, or whatever and let them sit for a time before serving.

Red and pink pepper are, of course, from entirely unrelated plants, the former from **chiles** and the latter from the **pink peppercorn** tree.

long pepper (*Piper longum* or *P. officinarum*): Long pepper is the fruit spike of either *P. longum*, native to India, or *P. officinarum*, indigenous to Indonesia. In both cases, long pepper is a creeping as opposed to a climbing vine, with erect fruiting branches about an inch or so in length which bear the catkins, or spikes, that are long pepper. *P. officinarum* has larger spikes of greater pungency than *P. longum.* The spikes are harvested when fully grown but while still unripe. After gathering, they are immediately sun dried or oven dried so that they do not mold. These dried long-pepper spikes look like tiny elongated, gray-brown pine cones.

Both long pepper and black pepper contain the alkaloid piperine, which is principally responsible for their bite on the tongue; but long pepper has a sweeter perfume, which resembles a combination of the aromas of black pepper and **ginger**. Its taste is also sweeter, and it numbs the mouth like **clove**. The larger spikes of long pepper can be hard to bite into, but an electric coffee grinder will handle them easily. This pepper can be used anywhere that ordinary black or white pepper is called for, and its hot sweetness is especially good with any kind of pickle; just add a couple of whole spikes to the pickling liquid.

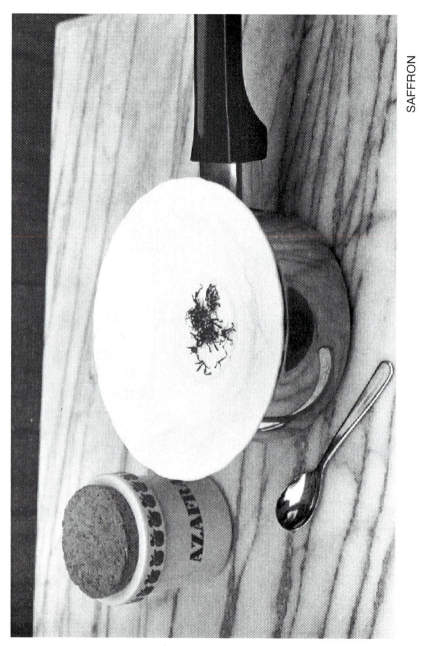

Long pepper was known in Europe centuries ago, imported from India to ancient Rome, and the first-century natural historian Pliny wrote that the Romans preferred long pepper to black. The use of long pepper continued into the Middle Ages in Europe, and it is called for in many medieval recipes. People who like to recreate medieval banquets can often find the long pepper they need among the bulk herbs in health food stores. It may be labeled "pippali," which is actually its ancient Sanskrit name and probably the source of our word "pepper."

cubeb pepper/cubeb/tailed pepper (*Piper cubeba*): The fruit of this member of the *Piper* genus resembles a peppercorn, but it is picked and dried with a little piece of stalk attached, giving it the name tailed pepper. The cubeb dries to a lighter color than the black pepper-corn, its color ranging from gray to brown. It is hot, aromatic and encourages salivation. Its taste is appealing, but more bitter than that of the other peppers described above, so it should be used more sparingly.

Native to the East Indies, cubeb pepper is enjoyed occasionally in Indonesian cooking, especially in curries and meat dishes. It may also put in an appearance in the multifarious Moroccan spice blend **ras el hanout**.

Cubeb pepper is available from specialty spice dealers. It must be stored carefully in an airtight container, as its flavorful volatile oils evaporate easily. Buy small amounts, and renew it frequently.

perilla/shiso leaf/beefsteak plant (*Perilla frutescens*): The beauti-ful jagged-edged leaves of this plant are a favorite in Japan for garnishing and flavoring food. Two varieties of perilla are used, called "green" and "red," although the color of the latter is actually a deep purple. The buds and seeds of green perilla also serve, like the leaves, as an edible garnish.

The green leaves are especially treasured for their perfume. This delicate aroma and the accompanying clean taste are a great comple-ment to tuna. Perilla is a standard garnish for sushi and sashimi; in fact it is so important in this context that, when the leaves are out of season or otherwise unavailable, their presence is often suggested by a strip of thin, jagged-edged green plastic served with the dishes as decoration.

Whole green perilla leaves may be deep-fried in a light tempura batter as an accompaniment to a Japanese tempura dinner.

Vietnamese cooks also use green perilla leaves, calling them *tia tô*, in salads and with grilled foods. A whole leaf wrapped in translucent rice paper, along with vegetables, meat, or shrimp, makes an attractive fresh Vietnamese spring roll, with a bright, fresh flavor.

The red variety of perilla leaves supplies the beautiful color to pickled Japanese plums, *umeboshi*, and to red pickled **ginger**.

There is no reason why this fine herb should be confined to Oriental cookery. It works extremely well when imported into other cuisines. One or two shredded perilla leaves in a fresh tuna salad are genuinely exciting. A chiffonade of large green perilla leaves (see Chopping Herbs in the "Culinary Practice" chapter), mixed with about twice as much finely shredded cabbage, makes an attractive bed for pork roasts and other cuts; this is a fabulous flavor combination. Mix fine perilla chiffonade into a light bread stuffing for strong-flavored fish to be baked or fried. Green perilla can also be used effectively in place of **basil** in various Italian recipes, especially with pasta—not as a substitute for basil, but to create an altogether new dish.

pink peppercorns (*Schinus terebinthifolius*): Pink peppercorns, sometimes called red peppercorns, made a splash in the culinary world a couple of decades ago as the signature spice of *nouvelle cuisine*. Their success may be due more to their pretty appearance on the plate than anything else. Similar to peppercorns in shape and size, these berries are nonetheless totally unrelated to **pepper**. They are the fruits of a small tree native to South America, but now grown commercially on the island of Réunion, in the Indian Ocean, for export mainly to France. Most of the pink peppercorns that we buy in the United States come to us via France.

Pink peppercorns are not peppery. Their flavor is more resinous than pungent. For looks and an interesting flavor, sprinkle a handful of pink peppercorns on top of grilled tuna, or toss a few into a tuna salad. They are also good in chicken salad or pasta salad. These fruits are simply used whole and uncooked. In South America, they have traditionally been used as a seasoning, a medicine, and the basis of an intoxicating drink.

These pink berries form part of an attractive peppercorn mixture, along with black, white, and green peppercorns and whole **allspice** berries ("Jamaica pepper"); this mixture makes a very tasty steak *au poivre* when cracked or coarsely ground and patted thickly onto a steak before grilling.

Pink peppercorns should be used sparingly because their lingering resinous flavor tends to dominate the other ingredients in a dish. They are also reported to upset some stomachs severely. The rumor that pink peppercorns have been banned by the U.S. Food and Drug Administration is not correct. Perhaps the confusion derives from the fact that the plant—not the berries—has been outlawed in Dade County, Florida, where this pretty little tree was imported as an ornamental but became a prolific pest. Only the bees still love it there, producing a fine, tasty honey from its blossoms.

California is similarly infested with the related California pepper tree, *Schinus molle* (also called Peruvian pepper tree), whose berries are slightly larger but are used in the same way.

If you don't have pink peppercorns, try substituting green peppercorns (see **pepper**). They do not taste the same, but both spices are colorful, aromatic, and strong flavored without being very hot.

poppy seed/khus khus (*Papaver somniferum*): Many different species of poppy grow wild and in gardens around the world, but in the United States, federal law has made growing *P. somniferum* strictly illegal. Yes, this is the poppy from which opium is made, but the seeds have no narcotic properties whatsoever—although it has been reported that poppy seeds can sometimes cause a person to fail a drug test, probably because of miniscule amounts of latex adhering to the seed coat.

The two main varieties of this plant are called black and white, the former producing slate-blue seeds and the latter cream-colored poppy seeds. Rarer are brown and gray seeds sometimes found in the Middle East and Ukraine. The dark blue-black seeds, imported from Holland and sold in standard spice jars by most spice companies, are the most common in the West. These seeds, although tiny, are plump by poppy seed standards; they make the lighter ones appear minute! The white variety is grown and used in India and other parts of Asia, and is usually available in Indian shops in the United States. Dark poppy seeds are stronger in taste than the light ones, while white seeds are mellower and slightly sweeter; both varieties are nutty, crunchy, and rich.

A magnifying glass reveals that poppy seeds are kidney shaped like miniature beans. They are loaded with a pale fixed oil, nearly odorless and with a mild almondlike taste, that is sometimes used in Europe for salads. Like the seeds from which it is expressed, the oil is free of narcotic properties.

These seeds are a popular way to give texture and mouthfeel to rolls, bagels, pretzels, breads, and cakes, either as a topping or incorporated into the batter. In Germany and Eastern Europe, generous quantities are used in fillings for strudels, hamantaschen, and other pastries. Curries thickened and enriched with poppy seeds are particularly popular in northern India.

Poppy seed dressing is well-known and much liked on green salads. A homemade poppy seed vinaigrette drizzled over slices of your favorite fruits makes an intriguing buffet dish; try making it with **lemon** juice in lieu of vinegar. A sprinkling of poppy seed is welcome in a coleslaw. Two to four teaspoons of this spice, added late in the cooking, will give substance and texture to a vegetable stir-fry or a medley of stewed vegetables.

The seeds are also excellent in a melted-butter sauce to season a side dish of noodles, rice, or boiled potatoes. This simple sauce is ideal for grilled fish. To make the butter sauce, melt the butter with only a small quantity of seeds—a scant half teaspoon with a tablespoon of butter makes an amount suitable for half a cup of (uncooked) rice. Stir them together over a low to medium flame just until the butter begins to brown; at this point, the seeds will be sufficiently cooked, and the butter will have a more interesting flavor. Immediately pour the poppy seed butter over the dish and serve.

Poppy seeds can be a bit stubborn about giving up their flavor. Without a little encouragement from the cook, they have a tendency to just lie there. You can coax the best nutty texture and flavor from them by toasting, soaking, or grinding. With any of these three methods, heat should be applied at some point in making the recipe. The first method—toasting—yields lively, crunchy poppy seeds ideally suited for salad dressings or slaw. Toast the seeds in a dry skillet over low heat for just half a minute, stirring or shaking the pan almost continuously. Pay attention, because it's easy to burn the seeds. Don't be deceived by color: even the white seeds, by the time they turn brown, are well past burned! Watch the second hand on the clock, and stop at thirty seconds. After toasting, pour the seeds out of the skillet, let them cool, and proceed with your recipe.

Soaking is a great way to get the most from poppy seeds for cakes and breads, whether they are baked in the batter or sprinkled on top. To make the seeds softer, sweeter, and more full-flavored, put them in a small saucepan with an equal quantity of milk. Bring to a boil, then remove from the heat, cover the pan with a lid, and let them sit for about an hour. The seeds will swell, soaking up much of the liquid. Replacing the milk with water will give equally good results. Strain through your finest strainer, or pour off the excess liquid and drain the seeds on a paper towel. Mix the seeds into cake or muffin batter, or "glue" them onto bread, rolls, or pretzel dough with a wash of beaten egg or dilute honey.

Grinding the seeds is necessary for poppy seed fillings, and it can be a challenge. Pounding in a mortar is a time-honored, if laborious, solution to the problem; this is somewhat easier when the seeds have been toasted. Be sure to use a ceramic or metal mortar because a wooden mortar would soak up the oil from the seeds, which would eventually turn rancid.

Another grinding technique was devised by clever cooks in Central Europe: a special poppy seed mill, rather like a small meat grinder, clamped to the table and hand cranked. These can be found, with diligent searching, in specialty food shops or at a spice dealer's shop. Because of the difference in scale, most poppy seeds will pass through an ordinary meat grinder completely unscathed. Food processors and blenders are also designed for larger prey than minuscule poppy seeds, but a coffee grinder will sometimes work well enough; don't overdo it, or the copious oil will separate out. Properly ground poppy seeds are voluminous, light, and fluffy, with no visible trace of the oil.

Indian cooks briefly fry their poppy seeds in hot oil with other spices before adding them to curries. Note that only the white seeds are used in India, because they do not spoil the color of the gravy.

For a poppy seed pastry filling, the crushed seeds are mixed with sugar or, more often, honey and a little water, then seasoned with compatible ingredients such as **cinnamon, vanilla, lemon** zest, chopped nuts, a little **chocolate**, or butter. These fillings are rich! Canned poppy seed fillings are sometimes available in the supermarkets; these generally consist of crushed poppy seeds and honey, leaving it to you to add any other seasonings you may desire.

Poppy seed fillings are spread on the unbaked doughs for strudel, coffee cake, yeast rolls, or whatever. The dough is either rolled up or a layer of additional dough is added. They are then baked in the oven, where the heat not only cooks the dough, but brings out the very best in the seeds.

Store poppy seeds in the refrigerator, in a lidded glass jar sealed inside a plastic bag. Theoretically, the seeds can be kept for as long as six months in a cool, dry, dark cupboard, but we have no way of knowing how long they were stored before we purchased them. With such a high oil content—as much as 50 percent—poppy seeds are subject to rancidity, an oxidation of the oil that not only spoils its aroma and taste, but also makes it potentially dangerous to ingest.

Sesame seeds can substitute for poppy seeds as a topping for breads, but for any other recipe these seeds behave too differently to replace each other.

R **rose/rose petal/dried rosebud/rose hip/rose water** (*Rosa* species): What is more romantic than a rose? Rose petals and rosebuds make foods evocative and beautiful. Fresh roses laid around a cake or delicate petals strewn over a pudding are very affective. But a rose is not just a pretty face: it has flavor that can make foods alluring as well! Roses marry particularly well with sweets. Their mild flavor can be made more intense by the addition of a few drops of rose water (see below).

Of the many different species of rose, only those with a good fragrance are flavorful enough to eat. The damask rose (*R. damascena*), the apothecary rose (*R. gallica*), and the cabbage rose (*R. centifolia*) are popular species in the kitchen. If you are unsure about using your particular type of rose, simply nibble a petal to see if you like its flavor. Of course, any flower used for culinary purposes must be strictly organically grown, innocent of pesticides, fungicides, and other chemicals; roses from the florist are not suitable for cooking.

rose petals: Pluck the petals from the flower and rinse them in cool water; gently blot them dry. If the petals have a light-colored "heel" at the base, snip it out with scissors and discard. The easiest and most effective way to use rose petals is to scatter them over finished desserts such as rice pudding, **chocolate** cake, or fresh fruit, especially strawberries. The petals are also good and pretty atop chicken salad or seafood salads.

Rose petals may be steeped in white-wine vinegar, for use in dressing green salads; or buried in sugar crystals to make rose sugar to serve at elegant teas. Rose-petal jam is a favorite in Turkey and other parts of the Middle East, and it also makes a regal topping for such Occidental treats as pound cake, apple pie, or ice cream. Rose pouchong tea, produced in China, is a delicate combination of black tea and rose petals.

A very good imitation of rose flavor is produced by the rose-**scented geranium**. The leaves of this herb can be substituted for rose petals in all the ways discussed above.

dried rosebuds: Dried rosebuds have their own nostalgic, musty taste. They are usually ground into a seasoning powder. In Afghanistan, the powder is added to a refreshing cold drink of thinned yogurt, while in Iran a few whole petals grace a cold soup of yogurt, cucumber, and **mint**. Many Persian cooks sprinkle ground rosebuds over cooked chicken or rice dishes. Use this spice sparingly: half a teaspoonful will suffice for a whole roasted chicken.

The popular spice blends **ras el hanout** and **advieh** contain powdered rosebuds.

Whole dried rosebuds are excellent in tisanes, usually mixed with other herbs.

rose hips: The fruits of the rose are celebrated for their high vitamin-C content. They can be eaten fresh and uncooked, but they are generally quite tart, and the seeds and tiny bits from the blossom end are unpleasant in the mouth. Usually, the hips are stewed with water and sugar to make jam, or are simply puréed with sugar and served as a sprightly sauce for meats, especially game. In both cases, the soft cooked hips should be strained through a fine sieve or cheesecloth. Do not use aluminum or cast iron pots or utensils when working with rose hips.

Dried rose hips are lightly crushed and simmered in hot water to make a tasty, colorful tisane, to be drunk alone or added as a flavoring to regular tea. For brewing advice, see the Teas and Tisanes section in the "Culinary Practice" chapter. Use about half a tablespoon of hips for each cup, and be sure to strain out the crushed fruits and seeds before serving. Honey is a harmonious sweetener.

rose water: Rose water is a distilled product, but this is not to say that it is alcoholic. The process of distilling the essence from fragrant rose petals does not involve a fermented substance.

Rose water is intended to put a dish over the top in elegance and refined sensuality. Many Indian and Middle Eastern desserts are finished with a half teaspoon of rose water, and sugar syrup is frequently scented with this culinary perfume. Rose water is a popular flavor for *lokum*, the candy also called Turkish delight. The classic Persian combination of rose water and **saffron** is so luscious that it's almost embarrassing.

rosemary (*Rosmarinus officinalis*): A hardy shrub of the Mediterranean region, rosemary grows long, needlelike leaves on tough woody stems. It is one of the strongest herbs in use today. Its fragrance is profound, like incense in a church, and its taste is potent, too—pungent, bright, and piney. Often, sufficient flavor can be obtained by simply laying a fresh sprig of rosemary on top of a dish after it is cooked. This has the advantage of looking good as well: consider laying a long sprig atop roast lamb or a whole grilled fish, for example. In climates where rosemary bushes grow thick and tall, the twigs are cut to use as skewers for miniature shish kebabs.

Rosemary keeps its flavor well when dried, but the dried leaves are sometimes very tough and even sharp, and they should be strained from a sauce after it has cooked, or tied up in cheesecloth before they are added to a dish, for easy removal later on. Finely chopped or powdered leaves may be left in the dish. It is best to chop or grind rosemary just before using to avoid losing its aroma. Only a small quantity is needed: a teaspoonful of fresh leaves, or one-fourth teaspoonful of dried, should be sufficient for four servings of most dishes.

This herb is a particular favorite in Italy and the south of France. There is an affinity between rosemary and the classic Mediterranean flavors, such as **chive, lemon, tarragon, thyme**, and wine, which is why it is inevitably a part of **herbes de Provence** blends. And **garlic** adores rosemary! Make a rosemary-garlic marinade for lamb, mixing in a little lemon juice to tenderize the meat and olive oil to carry the other seasonings. A rich garlic soup will welcome a generous application of rosemary leaves; immerse a whole rosemary sprig in the soup while it is cooking, remove it at the end, and then garnish each bowl with a small sprig of rosemary for looks and aroma.

Rosemary and lemon combine happily with all kinds of poultry. Flavor olive oil with the juice of a lemon, along with a little zest, and a tablespoon or so of rosemary leaves for a wonderful marinade or basting liquid. Roasted birds can be stuffed simply with a cut-up lemon and a couple of rosemary sprigs in the cavity.

Rosemary accents lamb ragout, pork roasts, and veal chops. A compound butter of four tablespoons of softened unsalted butter, two teaspoons of chopped fresh rosemary leaves, and a pinch of salt is an ideal seasoning for red potatoes and other vegetables. Beautiful breads with rosemary leaves baked in make an excellent companion for all these dishes. For an unusual and interesting beverage, immerse a long sprig of rosemary in a pitcher of iced tea.

Branches of rosemary, soaked in water then laid on the hot coals, add interest to the smoky tastes of grilled meats. A sprig can also serve as a basting tool, as the meat cooks.

There are many varieties of this beloved herb, some with longer leaves, and some whose leaves are broader or greener; flowers may be blue, pink, or white; there are erect varieties of the plant as well as prostrate forms. The fragrances, and hence the flavors, vary also. If you are fortunate enough to have a choice, pick the one that pleases you.

S **safflower/bastard saffron/Mexican saffron/Turkish saffron** (*Carthamus tinctorius*): Look for safflower anywhere **saffron** is said to be cheap. The orangey-red and yellow petals of the flowers are a standard substitute for saffron—to which it has no relation whatsoever—and are used in just the same dishes as true saffron threads. But safflower is not to be scorned as a mere substitute: it has a fine yellow color and delicate flavor of its own, which have earned it a respected place in Mexican, Middle Eastern, Indian, Polish, and Russian-Jewish cooking.

To use safflower to color a dish, you have to use a lot; for example, a pilaf calling for a cup of uncooked rice needs about two heaping teaspoons of safflower petals. Another way to calculate it is to figure about five times as much safflower as you would need of thread (not powdered) saffron. The safflower must be soaked ahead of time; pour on just enough hot water to cover the petals and let them sit for two hours or longer.

The dry safflower petals, sprinkled on a dish after it has been cooked, can also make a very attractive accent.

Although traditionally it was the flower that was the esteemed part of this thistlelike plant, the seeds provide a vegetarian and kosher substitute for rennet for curdling milk in cheese making, and in recent culinary history safflower seeds have become popular as the source of a light-flavored cooking oil, the most unsaturated edible oil in common use.

saffron (*Crocus sativus*): Throughout human history, saffron has been *the* golden spice, not only because it is a potent source of a beautiful bright yellow color but also because a good deal of metallic gold is needed to purchase it. Saffron is the world's most expensive spice. Its high price is easily understood when you consider that each carefully cultivated, hand-harvested flower yields only three little threads of saffron.

These threads are the stigmas and styles, the female organs of the flower, and in the saffron crocus they are long, red, and incredibly aromatic. When you look at a single saffron thread, you can clearly see the broad stigma at the tip of its long style.

The shorter yellow stamens, or male parts of the flower, are not used—neither by us nor by the flower, since the saffron crocus has been cultivated for so long that it can no longer reproduce on its own and depends on human beings for its propagation. The specific epithet *sativus* in its scientific binomial means "cultivated," and you will not find the true saffron crocus growing wild.

This crocus species blooms in the autumn, producing a lovely purple flower followed in a few days by long grassy leaves. It is important not to confuse the autumn-flowering purple saffron crocus with the autumn-flowering purple colchicum, unfortunately sometimes called meadow saffron, because the colchicum is poisonous! But the colchicum flower contains no red threads, and it is marked by the distinctive habit of not acquiring its leaves until the spring; for this reason it has also been given the name "naked lady."

When buying saffron, look for blood-red threads with very few yellow threads among them; a brown color indicates age and loss of potency. The aroma should be strong. Keep it in an airtight container in a cool, dark place; do not refrigerate or freeze it. If your kitchen is very hot, move the spice to a chest or cupboard in another room. When stored properly, good-quality saffron should keep for three years or even longer.

All the major saffron-growing regions—Spain, Kashmir, and Iran—produce excellent saffron, but of course they have lower grades as well. In Spain, the various grades are called Sierra, Rio, and Mancha, in ascending order, with top-notch, hard-to-find Coupé representing the ultimate in quality. Kashmiri grades, from lowest to highest, are Lachha, Mogra, and Shahi. Kashmiri saffron is rarely found outside India, as the subcontinent generally uses all that Kashmir can produce and in some years even imports saffron from Spain. Exports of Iranian saffron are sporadic, but the quality of the product is usually very good. In Iran, top-quality saffron threads are sometimes tied together in a glorious aromatic red plume.

Threads are preferable to powdered saffron, because the grinding process inevitably releases some of the volatile essences; however, the big spice companies do an excellent job of controlling this loss. If you do use ground saffron, get it from a reputable source, because it is tempting and easy to adulterate the powder. To measure, use just the amount of powder that you can pick up with the tip of a sharp knife; this knife-tip of saffron should measure about one-thirty-second teaspoon, and is used in place of one-half teaspoon of threads.

Of course the right amount of saffron in a dish varies with the quality and age of the spice and according to your own taste, but as a rough guide, four servings require about half a teaspoon of threads, loosely packed, or approximately forty to forty-five whole threads. Note that in the flower the three styles are joined together at the bottom; if you find a complete tripartite style, each branch should be counted as one thread.

Saffron is very good with starches such as rice, pasta, potatoes, and breads. It is required by definition for *risotto milanese*, Indian biryanis, Cornish saffron buns, Russian *kulich* (an Easter cake), and the sweet, short-grain rice puddings of the Middle East called *zerde*, *sholezard*, or *zarda*. Saffron with fish, shellfish, or poultry creates famous dishes such as Spanish paella, French bouillabaisse, and Pennsylvania Dutch chicken pot pie. In fact, saffron and chicken have such an affinity for each other that chicken broth, perhaps touched up with a drop of yellow food coloring, can be used to suggest saffron, and conversely a pale saffron water makes a pricey but effective substitute for chicken broth. Less well known is saffron's ability to add a subtle richness to tomato sauces. In Arabia, saffron-flavored tea is so popular it's available in teabags in the supermarket; to make your own, toast and steep one-fourth teaspoon of saffron threads, then add the liquid to a four-cup pot of regular hot tea, freshly brewed. Almost all fruits blend well with this spice. And saffroned cream is divine; try making saffron ice cream!

To get the best color and flavor from your valuable saffron threads, they should be toasted, pulverized, and steeped as follows. First, toast the saffron lightly for three or four minutes on a plate set over a pan of boiling water. You can also toast the saffron in a warm (225°F) oven for a minute, but you must take great care not to burn it. After toasting, pulverize the threads with a pestle or with the back of a teaspoon, adding a tiny pinch of sugar or salt if you like, to help break up the threads. They should be fairly brittle after the toasting; if not, put the plate back over the water and toast them a minute longer. Next, soak the toasted, pulverized threads in a couple of tablespoons of hot (not boiling) water or other liquid from your recipe. To develop its flavor as well as its color, the saffron should steep for at least fifteen to twenty minutes, so you need to allow time for this procedure when you begin.

Finally, in order to maximize the fragrance, the saffron liquid (along with the bits of steeped saffron) should be added to your dish as late in the cooking as possible. Pour saffron liquid into a bouillabaisse about five minutes before the end of cooking. For biryanis and elegant Persian rice dishes, a portion of the cooked rice is spooned into the saffron water and these gilded grains are mixed back into the remainder of the rice; the result is an attractive variegated coloring. For paella and for *risotto milanese*, add the saffron liquid along with the water or broth so that it infuses the rice; *never* sauté the saffron threads with onions (or anything else) at the beginning of the recipe!

If you are using the saffron in a recipe that has no liquid ingredient at all, as for saffron butter cookies, then mix the pulverized threads with the sugar and proceed with the recipe. This saffron sugar can be used in any recipe in which heat is applied for a sufficient time; for example, apple pie can be given a Midas touch simply by sprinkling saffron sugar over the slices of fruit and proceeding as usual. Use one teaspoon of saffron threads for a whole pie. Be sure to gild the top crust of this King Midas Pie by brushing it with saffron-steeped milk about ten minutes before it comes out of the oven.

To make a saffron butter that will spread molten gold over a fish fillet as you bake or broil it, toast the threads and pulverize them very fine with a little **salt** (or use commercial ground saffron); then mix the saffron thoroughly into the butter with a fork, using one teaspoon of threads per quarter-pound (one stick) of butter. Form the butter into pats and refrigerate until needed. The saffron color will not develop fully until the butter melts.

If you wish to poach fresh fruit slices in a saffron sugar syrup or, as the Pennsylvania Dutch sometimes do, to boil potatoes in a pot of saffron water, then it is not necessary to steep the threads first: The cooking time in recipes such as these is sufficient to bring out saffron's qualities.

No "paella powder," "egg shade" or other food coloring will substitute for the earthy flavor of saffron. However, because of the cost of true saffron, it is quite common to find cooks who have

worked out a decent enough compromise by extending a little real
saffron with a colorant.

On the other hand, you *can* use too much saffron, creating an
unpleasantly bitter, medicinal effect. As always, take note of the
quality of the spice you have obtained, and adjust the quantities up
or down accordingly.

A long red saffron thread or two makes a stunning decoration when
laid out on a bed of golden saffron sauce, but be careful not to overdo
using the spice as a garnish, or the flavor will become too strong.

Several popular recipes combine saffron with **lemon** juice or
vinegar, as in saffron lemonade or saffron vinaigrette, but do take
care to balance this pairing of the serious, heady, profoundly earthy
flavor of saffron with these bright and chipper acidic tastes.

Turmeric, **safflower** petals, and **marigold** petals are time-honored
stand-ins for saffron. This is because all three yield a yellow color,
rather than for any similarity of taste or aroma. These spices are
sometimes disdained as "false saffron" or "bastard saffron," but
they have their own merits and are even preferred by those who are
accustomed to them. Whenever you find a cheap "saffron" on the
market, it is likely to be one of these three. Please consult those
entries for information on substitutions.

sage/garden sage/common sage (*Salvia officinalis*): The soft, peb-
ble-surfaced, gray-green leaves of sage have a distinctive aroma,
musky and camphorous at the same time, with an earthy taste. The
flavor can be domineering when sage is used in quantity.

Dried sage is offered in whole-leaf form, ground and "rubbed."
The rubbed leaves have been crumbled into a fluffy, cohesive,
velvety mass, while the ground leaves are reduced to a fine powder.
Use whole and rubbed sage when you want to emphasize the herb,
and the powder when you want it to blend into a mixture.

Recently, fresh sage leaves have been reliably available in many
supermarkets. They taste quite different from dried sage. Formerly,
the dried herb, which is stronger, muskier, and more astringent, was
much preferred, but today, with our increasing appreciation for
fresh ingredients, many cooks feel that the fresh leaves are superior.
Perhaps it is best to include both fresh and dried sage in our season-

ing repertory, using dried forms for cooking and fresh leaves for garnishing and mixing into uncooked dishes. If you would like to freshen a recipe that calls for dried sage, the following equivalences may be useful: for one-half teaspoon dried sage, substitute one tablespoon minced fresh leaves or four large whole leaves.

Sage has been traditionally associated with fatty meats, believed to make them more digestible and certainly to make them tastier. Thus, sage is standard in poultry stuffing, featured in a sage-and-onion stuffing for goose, and a dominant flavor in sage sausages. It is sprinkled over chicken livers, rubbed into pork roasts, pinned between veal and prosciutto for *saltimbocca,* served with eel, molded into cheeses, stirred into cream cheese, or used to make a versatile compound butter for all kinds of vegetables. And ground sage can be baked into biscuits to accompany any of these items. The same principle inspires other delightful uses for sage such as, say, slipping a couple of fresh sage leaves in with the lettuce in a bacon, lettuce, and tomato sandwich.

Sage stands up to **garlic** and onion, and complements **orange** and other citrus flavors. In small portions, it blends harmoniously with other native Mediterranean herbs such as **parsley**, **savory**, **rosemary**, and **thyme**.

Historically, Dalmatian sage from the eastern shores of the Adriatic Sea has been highly esteemed. Sage honey, made by bees which have sipped the nectar from the herbal blossoms, is also a specialty of that region.

Garden sage is an extremely variable species, and not all varieties are suitable for culinary use. Even among the culinary cultivars, some have nicer flavors than others. If you plan to grow your own sage, taste a leaf or two when selecting plants.

There are hundreds of other species of *Salvia.* Most of them are ornamental, but a few are well regarded in the kitchen, particularly pineapple sage (*S. rutilans*) and clary sage (*S. sclarea*). Both herbs are similar in flavor to garden sage, but they are milder and each has a nuance all its own.

The scent of pineapple sage has an uncanny resemblance to the eponymous fruit. Fresh leaves make excellent garnishes for fruit dishes and drinks of all kinds; they may also be chopped fine and added to fruit salads. The leaves go well with custards and other sweets. A tender young leaf of pineapple sage laid on a nibble of cheese makes a snack or appetizer as palatable as it is pretty.

Tall, decorative clary sage bears fragrant, heart-shaped leaves with pink edges. The leaves at the bottom of the stem are quite large, as much as six inches wide and eight inches long. They are virtually always used fresh. Their aroma has been compared to lavender, but it resembles even more the highly scented muscat grape and, in fact, clary sage has been used to make ersatz muscadel wine.

This herb may be used anywhere that garden sage is called for, but its milder, floral flavor makes it suitable for more delicate dishes as well, such as a simple omelette. A centuries-old recipe for the soft, fleshy clary sage leaves provides a gossamer treat: a few large leaves are dipped in a thin batter of flour and milk, fried quickly in butter, then sprinkled with powdered sugar or orange liqueur. Eat 'em while they're hot!

salam leaf/daun salam (*Syzygium polyanthum*): The aromatic leaves of this evergreen tree related to the clove are appreciated in the cooking of Indonesia and Malaysia. The leaves are oval with a pointed end, approximately two to three inches long. They are not strong-flavored, but they add a subtle herbal touch to soups, stews, and the popular dish *nasi goreng*—fried rice of a hundred variations.

Salam leaves are used in precisely the same way that **bay leaves** are used in many Western dishes; that is, dried leaves are preferred to fresh, a single whole leaf is used to season an entire pot, and the leaf is removed before serving.

Packages of dried salam leaves are often sold in Oriental food stores, but if you are unable to find them, substitute a **curry leaf**; this will not duplicate the flavor, but both leaves add an interesting, rather haunting quality to a dish.

salep/sahlab (*Orchis mascula*): Many books claim that **vanilla** is the only edible orchid, completely overlooking the pretty early purple orchid whose root gives us salep. This flower grows wild over a large part of Europe and Asia Minor, and its roots are gathered, strung on a string, and hung up to dry. Later, they are ground into a powder and mixed with milk and sugar to make a nourishing, warming drink to which the salep gives its flavor and thickening power. A sprinkling of **cinnamon** on top is a popular addition.

It is a mystery why salep is not better known today. Two and three hundred years ago, this beverage was a regular feature of life in London. At that time, salep sellers could be found on nearly every chilly street corner during the winter months. Today, in just the same way, salep is a wintertime favorite in Turkey and a few other places.

Powdered salep is a little tricky to use because of its extraordinary thickening power. Be sure to add cold milk to the powder rather than adding the powder to the milk; add it slowly and stir like mad until it is thoroughly mixed. About one and one-half teaspoons of salep powder will thicken and flavor two cups of milk quite nicely. Heat the mixture slowly over medium heat, stirring almost all the time.

Sometimes salep is available in Middle Eastern shops in boxes ready-mixed with sugar, powdered milk, and even cinnamon (or cinnamon flavor), so that all you need to do is add hot water or milk to one tablespoon of salep powder. This mix is quite tasty, although you may feel that you get better results if you mix it from scratch yourself.

Salep is also a standard ingredient in ice cream in the Middle East, making the frozen custard incredibly elastic. Ice cream vendors show off their yummy product by stretching it out in strings up to two and three feet long!

salt (*sodium chloride—NaCl*): Salt is essential to the human body, as necessary as water. The amount needed by an individual varies with temperature and activity, as well as other factors such as pregnancy or illness. A certain amount of salt is naturally present in the foods we eat. Among mammals, carnivores get all the salt they need from their meat diet; herbivores must find other sources. We omnivorous humans usually need some supplement of salt but, in fact, we often eat more than necessary because of its flavor and other desirable culinary properties. Modern processed foods generally have a very high salt content. There has been a great deal of concern about the connection between salt consumption and high blood pressure in some individuals, but it is not true that everyone needs or benefits from a low-salt diet; salt sensitivity is a very complex condition.

Nevertheless, the American national taste is moving away from saltiness, and some American tourists eating in Europe, even at fine restaurants, find the food too salty. Similarly, if you're following a recipe from an older American cookbook, you might want to reduce the salt. It is a good idea to start with only half the amount called for in the older recipes, then taste and add to your own preference. Please note that this does not apply to yeast breads, for which the amount of salt is critical. Salt affects both the activity of the yeast and the condition of the gluten in the flour. To make low-salt breads, find a recipe that has been developed for this purpose. Also, the preservative action of salt is important in pickling, and the prescribed quantity should not be reduced in pickle recipes.

Salt inhibits the growth of molds and bacteria, and many of our favorite foods are preserved or processed in salt or brine, such as olives, sauerkraut, bacon, corned beef, and salt fish, as well as Oriental soy sauces and fish sauces. See the **lemon** entry for a description of making preserved lemons with salt.

Salt should never be stored in any kind of metal container. Silver salt servers for the table need glass liners. Glazed earthenware salt cellars with wide mouths are ideal in the kitchen, as are wooden salt boxes with a hinged lid.

The use of salt in cooking requires careful attention. A little salt brings out the other flavors of a dish, and omitting it altogether can lead to a terrible blandness. Increased use of herbs, spices, and other seasonings compensates for the flavor of salt in a low-sodium diet, and they can also disguise the fact that too much salt is present. The addition of a little sugar is a favorite way of handling an oversalted dish, but a sour taste, say from a squirt of lemon juice or vinegar, is a useful remedy as well, and may be more suitable to the particular dish being rescued.

Salt should be added during cooking in order to penetrate the food; a sprinkling of salt from a shaker over the finished dish affects only its surface. However, this rule does not apply to microwave cooking, for which salt must be added at the end of the cooking to prevent the food from becoming tough.

Fried or scrambled eggs are more tender when cooked without salt; for the same reason, avoid cooking eggs in salted butter. Put salt in the sauce or filling for an omelette, rather than in the beaten eggs.

Salt sprinkled on raw vegetables will draw out their liquids by osmosis, a process known as "degorging," which is commonly used to eliminate the bitter juices in such vegetables as cucumber and eggplant; however, modern varieties of these vegetables have often had the bitterness bred out of them, making this step unnecessary. After degorging, rinse the vegetables well to remove the excess salt. Similarly, sprinkling salt on meats draws out the juices, and some cooks prefer to salt meats at the end of cooking to keep them moist. However, this is a controversial point, and others prefer the flavor advantage of adding the salt during cooking.

Osmosis again requires that vegetables be boiled in salted water or other liquid, to keep their minerals from dissolving into the cooking water and being lost. This makes them more nutritious and helps to keep them crisp. On the other hand, cooking pulses such as dried beans, dried peas, and lentils in salted water makes them tough, and salt should be added only after the pulses have cooked enough to become tender.

Salt can be used rather spectacularly to encase a whole fish *en croûte de sel*. Choose a sole or other white fish for this French specialty. Chinese chefs are also expert at baking fish and poultry in this way. Lay the fish (or chicken) on a deep bed of salt in a baking dish, then cover it completely with more salt. Put a lid on the dish and bake in a hot oven (450°F). When it is done, the salt will have formed a hard crust; crack it open at the table, and the steamy fish emerges moist and succulent but—miraculously—not oversalty.

table salt/common salt: Table salt is usually offered in a fine grind, with a variety of anticaking substances added so that it pours in rainy weather; sometimes sugar is also added as a stabilizer. Since 1924, table salt in the United States has customarily been fortified to supply our nutritional need for iodine, a health measure that has effectively eliminated goiter in the nation. For those who do not like the taste of iodine in their food or who do not wish additional quantities of iodine (perhaps because they feel they are getting a sufficient amount in their city's water supply), iodine-free table salt is also available on the market.

kosher salt: In Jewish culinary traditions, salt is important for its role in making meat kosher. Coarse grains of salt draw out any of the prohibited blood that may remain in the meat.

If you use kosher salt for cooking, you will need to adjust the measurements in most recipes, because these coarse grains do not fill a measuring spoon as completely as fine grains do. (You can reduce the size of the grains with a salt grinder if you like.) The lack of additives in kosher salt also affects the saltiness of a spoonful, so a little experimentation is necessary until you develop a feel for using it.

Coarse kosher salt is ideal for topping pretzels and breads. Use a beaten egg yolk as the "glue" to attach the salt crystals just before baking. These large grains of salt are also good for salting the rims of cocktail glasses for margaritas or bloody marys: Pour a layer of salt into a saucer, rub the rim of an empty glass with a wedge of fresh lime, then touch the salt with the moistened rim.

Because it usually contains no anticaking substances, kosher salt is your best choice for pickling, when you want a clear brine—but do read the label to be certain that there are no additives, or buy a salt labeled expressly for pickling.

sea salt: While most table salt and kosher salt is mined from underground deposits, sea salt is evaporated from seawater or from salty springs. Traces of many different minerals combine to give it its taste and sometimes a bit of color as well, which varies depending on where it comes from, how it is harvested, and how it is dried. Sun-dried salt is hard to find, but particularly delicious. From around the world, there are rare and exquisite sea salts that well repay their price and the effort to find them in their effect on the flavor of foods.

Sea salt is available in coarse or fine grinds, and usually contains no additives—but check the label on each brand to be sure.

rock salt: Salt both raises the boiling point of water and lowers its freezing point. Rock salt, which usually has not been refined enough to cook with, is used to lower the freezing point of the icy slush around an ice cream freezer—just as it lowers the temperature at which the water on streets and sidewalks will freeze.

black salt/kala nimak: Dark-colored potassium chloride (KCl) has a slightly salty taste along with an accompanying bitterness. It is used in many salt substitutes designed for those who wish to avoid sodium. Commercial salt substitutes should not be used in cooking, but may be added to the finished dish according to taste.

Naturally occurring black salt is popular in India for its earthy taste and somewhat sulfurous aroma; it is sprinkled over fruits and salads, and is part of **chaat masala**, a popular seasoning blend for snacks. On occasion, black salt substitutes for sodium chloride for religious reasons.

sour salt/lemon salt: These are names for citric acid. See the Citrus section in the "Culinary Practice" chapter for more information.

Hawaiian salt/red salt/alae/'aleae: This flavoring is unique to the Hawaiian islands. It is important at traditional feasts where each guest is offered a small amount to add gradually to his foods as he delicately eats with his hands; the fingers are first dipped into the seasoned salt, then pick up the food. Hawaiian salt is made by mixing sea salt with a small amount of red ferruginous ocher, which is gathered from certain clay deposits found on a few of the Hawaiian islands.

sassafras (*Sassafras albidum*): Although many generations of Americans have enjoyed the fragrant root bark of this native tree as a flavoring for medicines, chewing gum, and both homemade and commercial soft drinks, this bark has now been relegated to potpourris and other nonculinary uses. The U.S. Food and Drug Administration has determined that sassafras bark is unsafe because of its high content of safrole, a precarcinogen, and its lovely rich flavor shall never darken our root beers again.

sassafras leaves/filé/gumbo filé: The variably shaped leaves of the sassafras tree are still available as a foodstuff, however. They are important as a thickener and seasoning in Creole gumbos, those popular seafood (and/or meat) and vegetable stews. The powdered dried sassafras leaves are known as filé or gumbo filé and the dish seasoned with the leaves is filé gumbo.

"Gumbo" comes from an African word for okra. When Africans reached America, they discovered that sassafras leaves contain an abundance of mucilage, as does okra; thus the leaves make a fine substitute for the vegetable, albeit with a different flavor. The flavor of the leaves is herbal and astringent. Filé is always an alternative to okra, and no gumbo should contain both.

The filé powder should be stirred into the gumbo at the last moment, after which the dish should not be boiled, or else it will become impossibly stringy. Use about one-half teaspoon filé per large serving. The filé powder can be set on the table for those who want to add a little more of the flavor.

If filé is cooked in a cast-iron pot, the gumbo will turn an unattractive dark color.

Sassafras leaves can also lend flavor to a ham. Lay whole leaves all over the meat, then bake it tightly covered; before serving, remove the leaves and decorate the ham with fruits such as candied cherries or fresh kumquats. Glaze with honey or a sweet sauce.

savory (*Satureja* species): Two species of savory are used in the kitchen—**summer savory** (*S. hortensis*), a cultivated annual with pale pink flowers and sparse, bright-green leaves; and **winter savory/mountain savory** (*S. montana*), an evergreen perennial with woody stems, profuse flowers, and narrow leaves of a darker green. While it's not exactly true that each herb has its season, summer for the one and winter for the other, you are more likely to gather *S. hortensis* while it's available—essentially, the summertime—and to turn to *S. montana* in the winter, when the summer savory has set seed and withered away. Select the smaller, younger leaves of winter savory, when possible.

These two Mediterranean herbs have in common a balsamic, peppery, literally mouthwatering flavor, reminiscent of **thyme**, and either one may be used when savory is called for. But that is not to say they taste just the same. The leaves of winter savory are tougher, more pungent, and have a stronger odor, which is just on the verge of smelling skunky. Summer savory leaves are more tender, with a lighter scent and milder taste, and can be used in larger quantities than those of their more robust sister herb.

In Germany, savory is known as *Bohnenkraut*, the "bean herb," because of the magic it does with beans and legumes of all kinds. A pinch of dried savory or two pinches of chopped, fresh leaves will enchant a simple pot of fresh green beans. Savory wakes up lentil dishes, and is de rigueur with favas. It makes creamed peas zestful.

Savory is especially appreciated in the south of France, where it grows wild over the hills, and is called *sarriette*. The same name is given to a soft goat cheese that is seasoned with savory. And it is an important ingredient of the popular blend **herbes de Provence**.

This delightful, peppy flavor is rather unjustly neglected in America, and could be used to advantage in herbal salad dressings, vegetable juices, chowders, tomato soup, and tomato sauces. Add just a little to three-bean salad or season succotash with it. Savory not only enhances the taste of Brussels sprouts, cabbages, and turnips, but a sprig of leaves, or half a teaspoonful of the dried herb, added to a quart pot of water, ameliorates the odors of those vegetables as they cook.

Combine whole leaves with an assortment of other fresh herbs, such as **chives**, **marjoram**, **parsley**, or **tarragon**, to make a flavorful bed for baked or poached fish. These resilient leaves are an excellent addition to meat loaf and poultry stuffing, and an ideal choice for herb bread. Savory flatters fruits; use a sprig with pickled peaches or stewed prunes, or infuse half a teaspoon of dried herb in the syrup.

Substitute a combination of two parts thyme and one part oregano for savory, or use a sprig of savory in a **bouquet garni** in lieu of thyme. Dried winter savory makes an excellent **za'atar**.

scented geranium/scented pelargonium (*Pelargonium* species): Unlike the common garden geraniums, which are grown for their showy flowers, the scenteds are grown for their beautifully sculpted, fragrant leaves, which vary widely in size, shape, color, and aroma. Botanists are unsure why, but these plants exercise the uncanny ability to mimic the scent of various fruits, spices, nuts, and other seasonings. There are rose geraniums, lemon geraniums, and mint geraniums, as well as others that smell like apple, coconut, lime, orange, strawberry, cinnamon, ginger, nutmeg, almond, lavender, musk, and on and on. Horticulturists fancy these natives of southern and eastern Africa, and their hybrids and cultivars are legion. The perfume industry uses the oils produced by the leaves of these herbs, and the food industry also uses them in gum, candy, and many other confections.

While all of the scented geraniums are technically edible, not all of them taste good. Cooks generally stick to the rose-scenteds, such as *Pelargonium graveolens*, lemon-scenteds such as *P. crispum*, and some of those with a minty fragrance, like the peppermint-scented *P. tomentosum*.

In the kitchen, scented geranium leaves offer almost as much visual pleasure as olfactory delight and taste. As a garnish, they are most attractive; choose a leaf whose scent complements what it garnishes. The lovely crinkly leaves of *P. crispum* are often floated in finger bowls, and this variety is sometimes called the finger-bowl geranium. The leaves may be dropped into glasses of iced tea, or frozen into ice cubes for fruit punch. Peppermint-scented leaves make an aromatic and refreshing tisane. (See Teas and Tisanes in the "Culinary Practice" chapter.)

A favorite use of rose geraniums is as a flavoring for cakes. Remove the stems from the leaves. Grease a suitable pan, and cover the bottom completely with the leaves. Pour in the batter and bake as usual. When the cake is done, garnish it with fresh leaves. A layer cake could be filled with scented-geranium jelly or iced with a frosting made of scented sugar. A pound cake might be topped with whipped cream infused with a rose-scented leaf.

Any of the culinarily inclined scented geranium leaves can be layered with sugar in a canister and allowed to sit for two weeks or longer, creating a subtly scented sugar. A few leaves may be added to fruit when making jelly or, even easier, buy a nice apple jelly, melt it gently, and pour it back in the jar to which a few whole scented leaves have been added. The flavor of the leaves will infuse the jelly within a few days. Cream or milk for custard, rice pudding, and other desserts may be scalded with a scented leaf or two; these leaves are strained out before the cream is used, and the finished dish is topped with fresh ones.

Scented geranium leaves add class as well as flavor to fruit dishes of all kinds; they are especially nice with berries.

The leaves are also infused in vinegar (see Flavored Vinegars in the "Flavor Combinations" chapter), which is then used in salad dressings. For quicker results, make a vinaigrette in the blender, buzzing in one or two scented geranium leaves at the beginning.

screwpine/pandanus (*Pandanus amaryllifolius*): We can only speculate that Europeans gave the inappropriate name "screwpine" to members of the *Pandanus* genus because the leaves grow in a whorl, or screw, and the fruit vaguely resembles a pine cone, although the plants are not related to the pines. The long, slender, pointed leaves—shaped like the blade of a sword—have a long-lasting, warm, earthy aroma.

These aromatic leaves are esteemed in the cuisines of Southeast Asia and India. Called *daun pandan* in Malaysia and Indonesia, *bai toey* in Thailand, *la dua* in Vietnam and *rampee* in Sri Lanka, they are used to perfume, flavor, and sometimes to color foods of many kinds.

Oriental stores often carry fresh leaves. Each leaf has a shiny side and a matte side; if both sides have a dull finish, this is a sign that the leaf is no longer fresh. Frozen leaves are also sometimes available, as are dried leaves; the dried ones are far inferior to fresh or frozen.

The easiest way to enjoy screwpine is simply to fold up a leaf and add it to the pot when you make rice. Use one leaf for about a cup of uncooked rice; fold it up and add it at the stage when you cover the pot and keep the lid on, so that the fragrance is not lost. The leaf is removed before serving, but its scent lingers, along with a delicate grassy, floral taste. Leaves or pieces of leaves are similarly added to curries; this is especially popular in Sri Lanka. The pieces may be left in the dish for the diners to push aside but, like **bay leaves**, they are too tough and fibrous to be eaten.

A popular Thai dish features cubes of marinated chicken meat wrapped in screwpine leaves. Since the leaves are so slender, each cube requires two leaves, wrapped around at right angles to each other. The wrapped cubes are fastened with a toothpick and quickly stir-fried.

Screwpine leaves are a favorite seasoning for sweets. They marry well with **coconut**, and are also used with rice, tapioca, or other puddings. The many sticky-rice treats so popular in Southeast Asia, whether sweet or savory, are usually enhanced with screwpine. For these purposes, the juice of the leaves is extracted to give an unusual light green color to the food in addition to its fragrance and flavor. An important example of this sort of treat is the Vietnamese New Year's specialty called *bánh chúng*; these pale green packages of meat and spice are ubiquitous at Tet.

For color, the leaves need to be bruised. A blender works well for this purpose. Add about one-quarter cup of water per leaf, and give it a few quick buzzes. Be sure to strain it well: those tough fibers can tie your tonsils in knots! Or you might prefer the traditional trick of tying a lightly pounded leaf into a loose knot to keep the fibers from escaping; the knot is tossed into a pot of water and heated. When the color and flavor are right, the knot is removed. Some recipes call for this liquid to be cooked with sugar to make a syrup.

A concentrated extract of the leaves is available in cans or bottles. This should be diluted with water but, because there is so much variation in the products, the exact ratio of dilution will have to be determined by experimentation. Resist the impulse to add extra juice in an attempt to get a brighter color, or the bitter grassy component of its flavor will become overwhelming. In the absence of screwpine leaves or extract, a little green food coloring will give authenticity to the *look* of your Southeast Asian dishes.

kewra water/kewda water/karwa essence (*Pandanus fasciularis*): The flower of another species of screwpine is distilled for its essence to make a culinary perfume similar to **rose** water and **orange** flower water. This soft, sweet scent is prized in India for ice cream, puddings, and other sweets, and it highlights the most elegant biryanis. Just a few drops in a small quantity of water or milk are sprinkled over the dish at the last moment. Sometimes both kewra water and rose water are extravagantly used together on a single dish! But do be careful with quantities; only a breath of each fragrance should be present.

Refreshing fruit drinks are often enchanted with kewra water. This essence transforms a simple lemonade into something exotic. One-quarter teaspoon of essence per tall glass is plenty. You may occasionally find kewra-scented fruit syrups, such as passion fruit, meant for mixing with ice and water.

Small bottles of crystal-clear kewra water are easy to find in Indian shops. Rose water is always an excellent substitute, and orange flower water is also appropriate. In the absence of all three flower essences, use a little **vanilla** extract. Because of its color, vanilla is fine for adding to most drinks and sweets, but it is not acceptable for sprinkling over a biryani or other light-colored dishes. For these dishes, add dilute vanilla extract in very small amounts.

sesame/sesame seed/benne seed/ til/ ajonjoli /gingelly/jinjelan (*Sesamum indicum*): This tall annual herb has an interesting strategy for distributing its seeds. As they ripen, the seedpods dry more and more until the pods suddenly burst open explosively with a sharp sound, flinging their numerous seeds in all directions. This dehiscence has been commemorated in the magic words of Aladdin in *The Arabian Nights*: "Open, sesame!" (Today, horticulturalists have developed some nondehiscent varieties for more efficient harvesting.)

Sesame seeds have been valued by human beings almost since the beginnings of agriculture, and this Old World plant has become widespread in most of the warmer parts of the globe. The small, flat, droplet-shaped seeds contain a large amount of a nourishing, mild-flavored fixed oil, known for its resistance to rancidity. Sesame seeds are not aromatic (and are therefore not considered to be spices by some authors), but they have a rich, nutty taste and can season many different foods.

As the seeds spring from the pod, each one is encased in a hard, thin seed coat or hull that, depending on the variety of the plant, may be beige, brown, red, or black. No matter what the color of the hull, inside is a pearly white sesame seed. Usually, white sesame seeds are sold and used without their hulls, but some cooks prefer to leave the hulls on for the extra crunchiness and nutritional value that they contribute. (Watch out for the linguistic confusion that arises from "hulled"/"unhulled": does "hulled" mean they have retained their hulls, or—as with "pitted fruits"—does it mean that the hulls have been removed? Usage of these terms is not consistent.) Without their hulls, sesame seeds are sweeter and nuttier; with the hulls left on they are crunchier, stronger-flavored, and taste more like a grain than a nut.

Both hulled and unhulled seeds should be toasted before use, either in a dry skillet (see Toasting Spices in the "Culinary Practice" chapter) or in a medium-hot oven. Because sesame seeds have little fragrance, the toasting time needs to be judged by eye rather than by nose: remove the seeds from the hot pan when they turn a light brown. Of course, this method fails with black sesame seeds; these should simply be toasted over medium-high heat for two minutes or until they begin to release a stain of oil in the pan. The seeds can also be fried in butter to create a sauce for noodles or vegetables. If you are topping off a stir-fry with sesame seeds, a very quick stir in a hot wok at the end of the cooking will toast them nicely.

The uses of sesame seeds are as numerous as the seeds themselves. They famously grace the tops of baked goods of all kinds—breads, buns, bagels, cookies, and those rings called *semit* that sell faster than hotcakes on the streets of Turkey. The quantity of seeds used for these toppings ranges from a light sprinkling to a dense coating. Well-beaten egg white is often brushed over these baked goods to ensure that the seeds adhere. Since they will be baked along with the bread or whatever, they do not need to be toasted in advance. On the other hand, if the seeds are to be incorporated into the dough, as for the thin, crisp benne wafers of South Carolina and Georgia, then give them a very light toasting first.

Sesame toppings are good with savory foods as well. Try sesame-crusted fillets of fish such as sole, cod, or orange roughy. Cut the fillets thin enough—about one-half inch thick—so that they don't have to be turned. Then simply moisten the top of the fillet with teriyaki sauce or soy sauce and sprinkle generously with sesame seed before broiling. That's all there is to it!

Just about any cooked vegetable enjoys a sprinkling of sesame seeds; and a small amount can be added to any breading or breadcrumb topping.

From Asia come countless ingenious sesame treats. Some of these almost seem to be simply an easy way to eat these yummy seeds, as with the ubiquitous little blocks of pressed sesame seeds and honey.

More complicated sweets include beautiful sesame-studded balls of yam paste (about tennis ball size), filled with sweet red beans.

Black sesame seeds are particularly popular in China and Japan, although the lighter seeds are by no means neglected. The jet-black seeds make a dramatic garnish for noodles, egg dishes, and many other foods.

Vietnamese companies manufacture sesame-studded dried rice paper—sold as thin, brittle, glassy disks containing a dense scattering of black seeds—for wrapping fresh spring rolls and other foods. The disks are soaked in hot water until pliable, then blotted lightly and used immediately.

Sesame seeds of any color may used to make the Japanese table condiment *gomashio*. This tasty combination of toasted seeds coarsely ground with sea **salt** also has an important place in macrobiotic cuisine.

tahini/tahine/tahina paste: This thick, oily, stiff paste, much used in the Middle East, is obtained by grinding whole, untoasted sesame seeds, resulting in a light beige tahini that is the basis of popular dips such as hummus (with chick peas) and baba ghanouj (with grilled eggplant, **lemon** juice, and olive oil). Thinned tahini, brightened with lemon juice and **garlic**, makes an excellent sauce for falafel or any kind of roasted vegetable. Whole vats of tahini are dedicated to the celebrated confection halvah (halwa, helva), made in blocks of assorted flavors by professionals.

Tahini keeps well in a closed container stored at room temperature or below. The oil has a tendency to separate out, so it should be stirred well before using.

sesame oil: Whether oil is expressed from toasted or untoasted sesame seeds determines its color, flavor, and uses. In south India and Burma, the pale oil from untoasted seeds is used for cooking, and in the West, this mild-flavored oil goes into margarines and salad oils and dressings. Sesame oil has a low burning point, so it is often combined with some other vegetable oil. In China and other parts of Asia, the sesame seeds are toasted before they are pressed, and the oil is a caramel brown color, with a strong, nutty taste. This oil is not used for cooking, but is added to a finished dish in small quantities as a seasoning.

soapwort/bouncing bet/'irq al-halawa/shersh al-halawi/helva wood/bois de Panama (*Saponaria officinalis*): The pale, woody root and stem of this plant, usually with bark attached, yield a sudsy, soapy extract that has often been often used in detergents and shampoos; but it has also found a few culinary uses, especially in the Middle East. In America, soapwort is generally found in health food stores, often in powdered form; pieces of root and bark are preferable when you can find them.

This obscure seasoning serves as an emulsifier for commercial tahini halvah, a favorite Middle Eastern sweet composed of crushed **sesame** seeds, sugar, and flavorings. It also forms the basis of a frostinglike, creamy dip, called *naatif*, for the nut-filled pastries *karabij*, which are popular in Lebanon and Syria. (In a pinch, marshmallow spread can be substituted for naatif; just flavor it to taste with a few drops of **lemon** juice and **orange**-flower water.)

To make either halvah or naatif, first begin with an extract of soapwort. Wash about one ounce of pieces of root or bark and bring them to a boil in one cup of water; then set the spice and liquid aside to steep for a day or longer. Finally, boil it again and reduce the liquid to one-quarter cup after the soapwort pieces have been strained out.

If you have powdered soapwort, whisk one tablespoon into one-third cup water until it lathers nicely, being careful not to inhale the powder as you measure and mix (see below). Cover the bowl and set it aside for an hour or more. Then strain the liquid through a piece of cheesecloth held in a sieve; wring out the cheesecloth lightly, and your extract is ready to use.

Put one cup of sugar and one-third cup water in a saucepan, and heat for a few minutes until the sugar has dissolved. Add one tablespoon each of lemon juice and orange-flower water or rose water (see the **orange** and **rose** entries) and boil the syrup, uncovered, until the temperature reaches 250°F or until a small amount dropped into a cup of cold water forms a thread. Resist the impulse to stir throughout this cooking time! Remove the pan from the heat immediately when the right stage has been reached.

Finally, in a bowl large enough to allow for a doubling of the volume, drizzle the hot sugar syrup slowly into the soapwort extract, beating vigorously and continuously until the mixture is thick, shiny, and bright white. Stiff egg whites are sometimes added at this point, but they are neither necessary nor recommended, because of the risk of salmonella from uncooked eggs.

The sweet, white, thick and creamy sauce thus produced is naatif, the exquisite accompaniment for karabij pastries. In Turkey, it is also the raison d'etre of another popular sweetmeat, *kağıt helvası*, consisting of two paper-thin wafers, about ten inches across, glued together with this extraordinary sauce. To make tahini halvah, the sweetened soapwort syrup is folded into the crushed sesame seeds, both to lighten the mixture and to keep the sesame oil from separating; the halvah is then further flavored, and molded into large blocks.

Soapwort root has an astringent, tealike flavor and a pleasant light fragrance that is fresh and—what else?—clean. For this reason, pieces of the wood are also used in Arabia to keep a broth fresh and sweet, much as **bay leaves** might be used in Europe or **mint** leaves in Mexico. Soapwort is frequently employed in this manner when making the famous Arabian fish and rice dish *sayadiyyah*, in which the rice cooks in the fish broth; the pieces of root are removed before serving.

Warning: Soapwort, like the other so-called "soap plants" such as soapweed (*Yucca glauca*), soapbark (*Quillaja saponaria*), and soap berries (*Sapindus* species), are relatively high in saponins, the substances that account for the long-lasting lather produced when parts of these plants are vigorously mixed with water. Saponins occur in varying amounts in many edible herbs and vegetables, but in large doses they can irritate the respiratory and digestive systems and can even function as a dangerously strong purgative. On ingestion, saponins can destroy red blood cells. Be very sparing in your use of soapwort; only a small amount is required to make a dish. If you have the powdered root, avoid inhaling deeply over it, as it is particularly distressing to the mucous membranes.

star anise/Chinese anise/badian (*Illicium verum*): This small evergreen Asian tree produces a beautiful aromatic fruit shaped like an eight-pointed star about an inch across. When the fruits are dried, each star point splits open to reveals a shiny, red-brown seed. The seeds smell and taste strongly of **anise**. The tree is botanically unrelated to the Mediterranean herb that produces anise seeds, but the two spices contain the same essential oil, anethole, and may be substituted for each other. Food companies in Europe and America often use star anise in the manufacture of baked goods, candies, and liqueurs.

The dried fruits are brown and woody, resembling a seed pod. They too are edible, although less intensely flavored than the seeds, and they have additional sour and bitter taste elements. Used together, seed and surrounding fruit make a complex, interesting flavoring. The effect is warm and sweet, as well as anisey.

Star anise is best bought unground, although the pretty stars are usually broken. You are lucky if you find a whole one—some spice dealers offer them at a special price—because they are stunning left whole in a fruit cup, bowl of punch, or atop a poached fish. Your guests can nibble their stars at the end of the meal as a digestive and breath sweetener.

Broken pieces of the spice will season a dish just as well as the whole spice; just count out eight points for each whole star specified in the recipe. In making one of the many red-cooked Chinese and Vietnamese dishes—simmered for long periods in soy sauce and spice—one star should suffice for about three servings, but use more if you like.

Star anise flavor is especially good with duck. Bits of the spice are sometimes included in the fuel burned to smoke fish and poultry. Powdered stars may be rubbed into pork before roasting it; use a teaspoon or two of spice for each pound of meat. One average star produces just over a teaspoon of powder when buzzed in an electric coffee grinder.

Star anise is commonly used to flavor Chinese tea eggs; these are hard-cooked eggs slowly simmered for several hours in black tea spiked with star anise, **cassia** sticks, and sometimes **cloves**. During the cooking, just after the eggs are hard, the shells are gently cracked all over to let the flavor and color penetrate in a network of fine lines. Serve tea eggs with a dip of pepper-salt (see **Szechwan peppercorn**).

Star anise is a key spice in *pho*—that redolent, hearty, steamy, spicy, herby, beefy noodle soup so loved by the Vietnamese. Malaysians put it in curries. Indians use it occasionally in biryani. As with anise seed, it is warm and inviting on carrots, parsnips, yams, and squashes.

Star anise is ground with other spices to make the celebrated Chinese **five-spice powder**.

Store this spice in an airtight container in a cool, dry place. Properly stored, it will last for years.

Substitute ground anise seed for powdered star anise, keeping in mind that the Oriental spice has the stronger flavor.

sumac/Sicilian sumac (*Rhus coriaria*): There are many species of sumac tree, and some of them are poisonous, but the fruits of this shrub or small tree are a beloved spice in the Mediterranean and western Asia. They are especially popular in the Middle East and Iran. The leaves are used for tanning leather and, for that reason, the tree is also known as tanner's sumac.

Sumac's taste is fruity, astringent, and sour, but not sharp. The small, moist, brick-red sumac berries are sun dried, then ground into a flaky, tacky powder; sometimes a little salt is added to facilitate the grinding of these moist berries. The powder is traditionally sprinkled over shish kebabs and grilled meats of all kinds. A sprinkling of sumac over pizza is an addictive modern adaptation of that use.

Ground sumac is also good with fish, however it has been cooked, and provides essential flavor and color to the Lebanese bread-salad *fattoush*. This spice is a basic ingredient, along with **thyme** and **sesame** seeds, in the Middle Eastern blend **za'atar**, and is part of a bright flavor trio of black **pepper**, dried **mint**, and sumac, meant for sprinkling over *mantı*. In another prevalent combination, mounds of thinly sliced onions with sumac and lots of flat-leafed **parsley** make a tasty accompaniment to meats and many other dishes.

Spinach pies, the little triangular pastries so well-known throughout the Middle East, have a savory filling of cooked spinach and, often, sumac. And sumac is featured in *musakhan*, a favorite Palestinian dish for special occasions, constructed of a round of Arab bread topped with chopped sweet onions that have slowly browned in olive oil, on which sits half a roasted chicken lavishly garnished with golden pine nuts and bright red sumac. Just the sight of it makes your mouth pucker and water in anticipation. Use lots of sumac for this dish: two tablespoons for each half chicken is not too much.

In general, sumac is not cooked, though it may visit the oven briefly, as when the little spinach-sumac pies are baked, or when the completed musakhan receives a final short reheating before it is served.

Sumac gradually turns brown as it ages. To keep its red hue longer, store it, tightly sealed, in the refrigerator. The older, darker sumac can still be used; it remains tart but much of its fruitiness vanishes with time.

This spice is easily found in Middle Eastern grocery shops or may be ordered from specialty spice dealers, but if you need a substitute, **lemon** juice is standard, even though the citrus juice has a sharper taste. Use a tablespoon of lemon juice for each teaspoon of sumac, or to taste.

Szechwan peppercorn/Sichuan pepper/Chinese pepper/fagara/anise pepper (*Zanthoxylum* species): The aromatic red-brown berries of small prickly ash trees have been an important flavor element in the cuisine of Szechwan and other regions of China since ancient times; different regions of the country use different species of *Zanthoxylum*.

Usually described as "hot," this spice produces a tingle on the tongue followed by a general numbness of the mouth, as if it *had been* hot. Westerners occasionally call these prickly ash berries descriptively "brown pepper" or "Chinese pepper," but of course they are "pepper" only in their function of spicing up food, and bear no other relation to the familiar black **pepper** (*Piper nigrum*). The Chinese call their native spice "fagara," while black pepper is often referred to as "foreign fagara." In the West, this spice is inexplicably sometimes called anise pepper, but it is not anise flavored. Rather, in addition to the tingle, the flavor has a citrus note.

The heat in Szechwan dishes is generated by **chiles**. The combination of chiles and Szechwan peppercorns—described as "*ma-la* flavor"—has long been popular in Chinese cuisine.

Whole Szechwan peppercorns, sizzled in oil then strained out, make a flavorful cooking oil for stir-frying. If chiles are added, then you have a ma-la oil. Store the spiced oil in the refrigerator, but do not try to keep it for more than about a week.

Szechwan peppercorns are an authentic ingredient in **five-spice powder**. Except in this form, the peppercorns are usually sold whole. When selecting this spice, check for fragrance and remember that a good brand will include relatively few of the brittle black seeds, which do not have the same pleasant flavor as the dry, red-brown fruits or pods. Szechwan peppercorns should be stored whole until needed, because they lose their flavor rapidly after being ground. Toast and grind them just before using. (See Toasting Spices in the "Culinary Practice" chapter.)

Pepper-salt, popular in Chinese cooking, is a fragrant combination of Szechwan peppercorns and **salt**, which is served as a dip for fried foods. Grind toasted Szechwan peppercorns well, mix thoroughly with twice as much fine salt, then heat the combination in the same dry skillet used for toasting the peppercorns. When the aroma wafts throughout the kitchen, remove the seasoned salt from the heat and cool. Black or white **pepper** is frequently added to this mixture as well. Use pepper-salt as a dip with Chinese tea eggs (see **star anise**).

There really is no substitute for the unique flavor of Szechwan peppercorns, but a mixture of equal parts of ground black pepper and medium-hot ground red pepper will mimic the tingle produced by this spice.

sansho/Japanese pepper (*Zanthoxylum piperitum*): This species of the prickly ash grows wild in Japan and is also cultivated in gardens. Its berries are somewhat lighter colored than most of the Szechwan peppercorns used in China. When ground, they yield a tan powder with a lovely clean aroma; this is the form in which sansho is usually sold.

A pinch of ground sansho powder is traditionally sprinkled on eel—only a little, or it's too spicy—and many other fatty foods. Ground sansho is usually offered as a condiment with *yakitori*: grilled skewered morsels of chicken or chicken skin, liver, or gizzards, with onions and other vegetables, which are served as hors d'oeuvres with drinks.

Sansho is part of the seven-spice blend **shichimi togarashi**, which is sprinkled over many hot noodle dishes.

The Japanese also use the fresh young leaves of the prickly ash, which they call *kinome*, both as a garnish and a seasoning. The leaves are milder than the peppercorns and emphasize the slightly lemony flavor of this spice. A graceful sprig of kinome topping a bowl of soup provides an aromatic garnish and symbolizes the arrival of spring. Sometimes the leaves are boiled with meats or fish to give a clean scent to the broth.

In northern Japan, a cake called *kiri-sansho* is made of ground sansho and flour. In the old tradition, the pestle used to grind the toasted pods was made of the wood of the same prickly ash tree, and also lent its flavor to the cake.

T **tamarind** (*Tamarindus indica*): The tamarind is a tropical tree, a legume, similar in appearance to an acacia. Its fat, lumpy seed pods, about an inch thick and ranging from four to eight inches long, contain up to a dozen hard shiny seeds encased in a dark brown, sticky, very fibrous pulp. The entire pod is covered with a crusty red-brown shell. The usual method of preparing tamarind for the market is simply to remove the outer shell, and press the pulp—seeds, fiber, and all—into flat cakes. These cakes should be stored in the cool, dry spice cupboard, and will keep indefinitely.

To use tamarind in the kitchen, break off a piece from the pressed cake and soak it in hot water for twenty minutes or longer. A good ratio is four times as much water as tamarind pulp; for example, one-half cup of hot water for approximately two tablespoons of pulp, or one cup of water for one-fourth cup of tamarind. These measurements are necessarily rough, because each piece of tamarind contains an undetermined amount of fiber and an uncounted number of seeds; besides, the incredible stickiness of the pulp makes precise measuring with spoons and cups quite impossible! As it soaks, break up the pulp and stir it around in the water. Finally, strain the liquid; press down on the pulp and scrape the underside of the sieve to get all the thick, golden-brown mass.

Tamarind pulp is sometimes also sold in a seedless block. This is used in exactly the same way as described above, except of course less of the pulp is needed to get the same intensity. Even though the seeds have been removed, the pulp must nevertheless be strained, because fibers are still present. Jars of tamarind concentrate, looking very much like pitch, are also available in many Oriental food shops. Dilute one-half teaspoon of concentrate with one-half cup of warm water to obtain the usual strength tamarind juice for recipes. Or add the concentrate directly to a dish as it cooks—use a mere teaspoon for six servings. Many cooks find the flavor of the concentrate inferior to that of the less-processed pressed pulp.

Whichever of the above methods is used, the resulting tamarind juice is acidic and tart, with a soft scent of sweet fruit and a taste that combines a tropical tang with a little smokiness. This juice is cooked with vegetables or used in sauce for meats, poultry, and fish. Iranian cooks traditionally cooked stuffed cabbage and other stuffed vegetables in this tamarind liquid. It is especially popular in the curries of south India. It also flavors chutneys and sambars. (See **sambar powder** in the "Flavor Combinations" chapter.)

Tamarind is an important souring agent in the cuisines of Southeast Asia. In fact, the Indonesian word for "sour," *asam*, is also the name of tamarind in that language. This spice is especially popular with fish, but there are many tamarind-laced meat and vegetable dishes as well.

Cooks in the Caribbean also make good use of this spice that grows so happily in that climate. It flavors both curries and barbecue sauces in the region. *Tamarindo* is popular in Latin America, too, for refreshing cold drinks.

Tamarind makes a regular appearance in Western cuisine as an ingredient of Worcestershire sauce and other commercial table sauces.

Strained tamarind juice, mixed with sugar and cold water—either sparkling or still—makes a refreshing drink, similar to lemonade. Spiced with **cumin**, a little **salt**, and **mint** leaves, it becomes the Indian beverage *jaljeera*. Tamarind juice can also be used to make a tantalizing sherbet; just use tamarind juice in lieu of **lemon** juice in your favorite recipe for lemon sherbet, ice, or sorbet, and reduce the sugar slightly. Make it sweet for a dessert, and not so sweet for an ideal palate cleanser between courses. Refrigerate any extra tamarind juice for as long as a week. It can be frozen for two or three months.

You may also find preserved tamarind candies or comfits, little balls of sweetened pulp rolled in sugar crystals. They are gratifying treats, but it is recommended to mumble them around in your mouth when you eat them in order to locate and extract that little bit of fiber that's sure to be there.

If you cannot get tamarind, substitute lemon or **lime** juice. Adding a fat pinch of brown sugar or a dollop of molasses helps to give the proper effect.

tarragon/French tarragon/true tarragon (*Artemisia dracunculus*): This delicate, slender-leafed herb is the darling of French chefs. Although it is a native of northern Central Asia, it has become indispensable to classic French cuisine. It is considered one of the **fines herbes**, and is also a part of **herbes de Provence**.

Tarragon's mild but distinctive flavor, sweet herbal with **anise** overtones, is essential in béarnaise sauce, ravigote sauce, and any dish suffixed *à l'estragon*. It is the perfect herb for lighter dishes such as poultry, seafood, and eggs. A sprinkling of finely chopped tarragon leaves makes an ordinary turkey gravy dramatic, and it brings elegance to lentils and brown rice. Incorporate a few fresh leaves in a green salad for a lively effect. Try tarragon over **orange** segments or with almost any dish involving oranges. Finely chopped tarragon makes an excellent compound butter to serve with baked fish. A popular Balkan yeast-cake recipe calls for an astonishing filling of fresh tarragon, sour cream, and butter-cookie crumbs, bound together with egg yolks and accented with **lemon** zest.

While dried tarragon is not as fine as fresh, it is still entirely usable if it is fully rehydrated. If your dish does not offer sufficient liquid or time, try soaking the dried leaves in advance in a little wine or dilute orange or lemon juice, depending on what is appropriate. The taste of fresh tarragon can also be preserved nicely in vinegar, which is then used in marinades, salad dressings, and prepared mustard. (See Flavored Vinegars in the "Flavor Combinations" chapter.)

It is difficult to grow tarragon during the winter anywhere. And this herb is not happy during any season at all in warm climates such as the southern United States; but fortunately, the sun-loving Mexican mint marigold makes an ideal substitute for tarragon (see **marigold**). Use the leaves of this herb just as you would use tarragon; if you find the marigold too strong, simply reduce the quantity slightly.

Russian tarragon, *A. dracunculoides*, which is similar in appearance but with a slightly broader leaf, has often been used in place of French tarragon. But this is an emergency measure only, as it is generally agreed that the taste of the "Russian" herb is decidedly inferior.

thyme/garden thyme/kitchen thyme (*Thymus vulgaris*): Of all the hundreds of species, varieties, and cultivars of thyme, this common garden variety is the one most used in the kitchen. Its small bright-green leaves are well supplied with the essential oil thymol, which contributes much to its characteristic potent, penetrating, and persistent flavor. Other thymes have slightly different chemical compositions and are named for other seasonings that they somewhat resemble in aroma and taste, such as orange thyme or oregano thyme. Caraway thyme, *T. herba-barona*, was traditionally used to season huge roast barons of beef (the double sirloins). Thymes hybridize notoriously, and you should sample a few leaves of any new plant to be sure of what you are getting.

One hybrid, lemon thyme (*Thymus* x *citriodorus*), has a definite, pleasant lemony fragrance, and offers a popular variation on thyme flavor. Lemon thyme can be used in place of garden thyme anywhere, but it is particularly well suited to fish, to fruits, and to onions and other sweet vegetables. This lemony herb can also substitute for **balm** or **lemon verbena** in equal quantities. It is not, however, a good substitute for the acidic **lemon** itself, nor is it an acceptable stand-in for **lemon grass**.

Thyme's flowering tops have the same flavor as the leaves, albeit in a milder version. They look pretty as a garnish or chopped and scattered lightly in a salad.

The sharp, warm-to-hot taste of thyme can overpower other flavors, and it should not be used to excess as a seasoning. Nevertheless, this versatile herb is indispensable in the well-seasoned kitchen.

Thyme has been called the secret of French cooking. It is important in ratatouille, cassoulet, pot-au-feu, and in nearly every meat dish in the classical French repertoire. It blends well with other Mediterranean herbs such as **basil**, **bay leaf**, **garlic**, **oregano**, and **parsley**.

Sheep, goats, and even snails that have fed on the wild thyme of the Mediterranean hills have acquired a better flavor from the herb. And bees are very fond of the nectar of thyme blossoms. The dark, redolent thyme honey from the herbs on Mount Hymettus near Athens has been esteemed since the time of the classical Greeks.

Thyme deepens the flavor of any dish braised in wine, from coq au vin to beef stew in red wine to poached pears for dessert; only a hint of the herb is needed—a pinch of dried thyme suffices for a cup of wine.

Thyme is a standard part of the classic **bouquet garni**, **herbes de Provence**, and packaged "mixed herbs" (see **mixed spices** in the "Flavor Combinations" chapter), and it's also an attractive option in many other herb blends. And of course thyme is the basis for the Middle Eastern blend **za'atar**. This versatile herb is surprisingly well matched with the sweet spices, **allspice**, **cinnamon**, **clove**, and **nutmeg**: a simple broth flavored with thyme and nutmeg makes a consummate soup.

Thyme is very popular in Caribbean cooking, where it is added to spicy sauces and is one of the popular **jerk seasonings**. This herb can stand up to the **chiles** in the dishes of that region as well.

Thyme butter is often melted over fish. A dusting with ground thyme really wakes up a fried egg. If you indulge in fried liver and onions, don't forget to scatter fresh thyme lightly over the finished dish. A legendary Balkan appetizer generally known as Albanian liver consists of small cubes of lamb's liver, lightly sauteed in olive oil, then finished off in a hot oven with a coating of red pepper flakes and thyme.

Thyme invigorates cheese dishes and cream of vegetable soups. A small amount will supply some backbone to the flavor of cobblers, compotes, and other dishes made with sweet fruits. Try a pat of thyme butter on a baked sweet potato. A pinch of dried thyme (or half a teaspoonful of fresh leaves) elevates any carrot recipe, *not* excluding carrot cake. Many herb breads contain this herb, among others. It is used to flavor olives, and its strong taste makes it the perfect herb for a tapenade of black olives, capers, and anchovies.

There is something of the taste of thyme in both **ajowan** and **savory**. Substitute savory for thyme in equal amounts, but if you use ajowan in place of thyme, reduce the quantity by about one-quarter.

turmeric/tumeric (*Curcuma longa*): This spice is derived from the rhizomes of a tropical plant related to **ginger**. Turmeric rhizomes, however, are generally smaller than ginger root, and the flesh under the brown skin is bright orange. Turmeric has a mild but rich, musky fragrance and a resonant, earthy and bitter taste that is warm but lacks the pungency of ginger. This complex, interesting flavor is extremely popular in Asia.

Asian markets in America sometimes sell packets of frozen pieces of turmeric and occasionally offer fresh rhizomes. These pieces can be peeled and added to the pickling bath for any kind of pickled vegetables. They will slowly tint the pickling liquid and add their own flavor, which complements both the vegetable and the **pickling spices**. Fresh or frozen pieces of rhizome can also be grated into the pickle.

More often, however, the rhizomes are "cured," that is, boiled with substances that bring out their color and aroma for cooking, then dried and ground to a bright gold powder. This spice is what makes **curry powder** yellow, and it provides the color in American hot dog **mustard**, the English relish piccalilli, and in margarine, cheeses, and beverages. Be careful when working with turmeric, because it is an excellent dyestuff for textiles—as well as for countertops, wooden spoons, fingernails, and almost anything else it comes into contact with! But these turmeric stains, while tenacious, are not indelible, and they will eventually wash out or scrub out.

Indian cooks believe that powdered turmeric should not be served uncooked; the unmistakable musky taste of raw turmeric can be detected under all the other spices of a curry. Many Indian recipes begin by sizzling the powder briefly in hot oil.

Because of the brilliant yellow color it imparts to food, turmeric is often used a substitute for **saffron**, even though the flavors of these two spices are entirely dissimilar. Turmeric is sometimes sold as Oriental saffron, Indian saffron, or even simply as saffron. This is not necessarily deception, as many people around the world do know the rhizome by that name.

Actually, mixing turmeric with saffron not only stretches the more expensive spice, but it also makes an intriguing flavor combination. This combination is frequently used in Morocco in the *harira* soup that is eaten to break the fast at Ramadan. The mix also makes a successful yellow rice, and often seasons couscous dishes. But do treat the spices in this mixed marriage properly: the turmeric profits from being fried, and the saffron threads should only be gently infused (see **saffron**). Keep the spices separate until they are added to the dish, each at its appropriate time. For example, for the yellow rice, the turmeric could go into hot oil at the beginning along with the grains of uncooked rice and, if you like, a little finely chopped onion; then the saffron liquid could be added later with the water or broth in which the rice will simmer until done.

A combination of turmeric and **coconut** milk lends an exotic flavor to soups or seafood dishes.

Some cookbook authors claim that a little turmeric in the cooking water will make green vegetables stay bright, while others caution that the spice will turn them a yucky gray. But experiments with broccoli, Brussels sprouts, collard greens, green beans, peas, and spinach, all cooked separately with and without turmeric in large quantities of boiling water for minimal times, reveal that—other than a slight yellow tinge—it hardly makes any discernable difference.

The amount of turmeric required varies with the dish, and usually your recipe will specify quantities, but if you need a general guideline, try measuring out about one-fourth teaspoon per serving.

V **vanilla** (*Vanilla planifolia*): The vanilla "bean," the fruit of a tropical American climbing orchid, has no fragrance or flavor when it is picked from the vine. By an amazing feat of human discovery and invention, those properties are induced by a months-long process of fermenting and aging the bean until crystals of natural vanillin develop throughout, giving it its perfume and savor. Many other components also contribute to the complex flavor of vanilla, which is why vanillin alone is not an adequate substitute for this seasoning.

Outside its native habitat, the vanilla orchid lacks natural pollinators and the flowers must be delicately pollinated by hand, one by one. It was in the second quarter of the nineteenth century that the technique of hand pollination became widely known, and only after that time was the vanilla orchid successfully cultivated in other tropical regions. Today it is grown commercially in several countries around the world, including the Comoro Islands, Costa Rica, Indonesia, Madagascar, Mexico, Tahiti and other Pacific islands, and the West Indies.

The flavor of the beans produced in different regions varies slightly, and each has its own devotees. Highly esteemed Bourbon vanilla is grown in Madagascar, Réunion, and the Comoros, and takes its name not from whisky but from the old French name for Réunion.

More dissimilar still is Tahitian vanilla, *V. tahitensis*, a species that was developed by horticulturalists early in the twentieth century, and is now grown on several islands in the Pacific. This vanilla is more richly endowed with a floral perfume, and less with vanillin. Some people prefer it; some don't. Experiment with vanilla beans from assorted regions to see which type you prefer.

The most common way of using vanilla as a flavoring is in the form of an extract. This is made by macerating the beans in a mixture of water and alcohol for a few days. Beyond this basic concept, however, manufacturers differ widely in the details of their procedures. The quality of the water and the alcohol varies, of course, as does their proportions; but the U.S. Food and Drug Administration requires that in order to be labeled "pure," vanilla extract must be at least 35 percent alcohol. Some manufacturers heat the liquid as the beans soak; others insist upon the slower cold-extraction process. After that initial step, some makers add sugar or corn syrup, colorings, glycerin, or other additives; some add nothing. Some then age the extract a bit, either in oak barrels or in stainless steel vats. Filtering is another optional step. The prices for the finished products also vary all over the charts. Clearly, it is worthwhile to try a range of extracts to find one that suits your purposes and your taste buds.

Mexico produces excellent vanilla beans and good quality extract, which is sold there in liquor stores, but unfortunately tourists often find an inferior, sometimes dangerous—but cheap!—vanilla-scented concoction containing synthetic vanillin and/or coumarin, which has been banned by the FDA as a food additive (see **woodruff**).

Another factor to consider when choosing a vanilla extract is the concentration of vanilla beans. The standard set by the U.S. government is 13.35 ounces of beans per gallon. If there is twice that proportion of vanilla beans, the extract is labeled "twofold"; threefold extract has 40.05 ounces of beans per gallon, and so on. Twentyfold is the limit, and this is made without alcohol. These more concentrated vanilla extracts are not found by the home cook in supermarkets, but are sold in industrial quantities to commercial bakers and confectioners.

Good vanilla beans should be a rich dark brown, slightly moist, somewhat flexible and, above all, fragrant. Inside each long, slender sheath are literally thousands of minuscule seeds, which are as flavorful as the pod containing them. Stored in a sealed container away from heat and light, a bean will retain its flavoring ability for years. (Similarly, a glass bottle of extract needs its lid screwed on tight, and storage in a cool, dark cupboard.)

There are various ways to extract the flavor from a whole vanilla bean. First, you can make a simple homemade extract by immersing a bean in brandy or vodka and letting it sit tightly sealed until needed. You can slit a piece of it lengthwise and scrape the tiny dark seeds into your dish—about one inch of scraped bean is roughly equivalent in strength to one teaspoon of pure vanilla extract. You may also steep the entire bean in warm milk to obtain its flavor; this can be done repeatedly with a single bean, rinsing and drying it thoroughly between uses, until the bean has lost its scent. Finally you might like to make vanilla sugar, as the Europeans do, by burying the bean in a canister of sugar; this is an excellent place to store a good vanilla bean.

Occasionally, vanilla powder comes on the market. This is the pulverized bean, sometimes mixed with sugar. It often aims to match the potency of vanilla extract, so that a teaspoonful of powder can be substituted for a teaspoon of extract; but this is not always the case. If you are unsure about the powder, add it in small amounts, then taste, until the desired intensity of flavor is reached.

Vanilla is as much a part of cake as is flour, and it mellows the taste of any sweet containing eggs, such as custards, creams, dessert soufflés, eggnog, flan, French toast, or Pavlova. It goes almost automatically into many pastries and cookies, and is a popular flavoring for yogurt and an indispensable partner with **chocolate**.

Vanilla is still America's favorite flavor for ice cream (chocolate ranks second). Variations on the vanilla ice cream theme include "whole vanilla bean," with little dark flecks which are the flavorful seeds of the plant, and French vanilla ice cream, enriched and colored a creamy yellow by the addition of egg yolks.

Most fruits respond happily to vanilla. A dollop of cream with vanilla, or a vanilla-scented sugar syrup, is good on fruit cup, stewed fruit, or pie. Add a dash of extract to fruit beverages, rum punch, or anything with rum. A little vanilla, along with cream and sugar, transforms a cup of coffee into dessert.

This rich flavoring is ideal in some savory dishes as well. Vanilla is luscious enough for lobster! Just add a few drops of extract to the melted butter you serve with it. Try it with crab, salmon, shrimp, and other seafood. Add a little vanilla to the glaze for a ham. A vanilla-flavored cream sauce with chopped pecans works magic on a grilled chicken breast. A hint of vanilla enhances mashed sweet potatoes or squash, and glazed carrots.

The high alcohol content of vanilla extract makes it highly volatile, so it is always added at the end of the cooking period to prevent the flavor being driven off by the heat. Whenever possible, as with custards and puddings, wait to add the extract until the food has cooled a bit.

Artificial vanillas only approximate the heady perfume and seductive flavor of this seasoning, and some develop an unpleasant aftertaste after a couple of days. For this reason, they are all best used in foods that will be quickly consumed, and avoided in any dish that is to be frozen for later serving.

W **wintergreen/teaberry/mountain tea/checkerberry** *(Gaultheria procumbens)*: This shy native North American herb is found growing wild in wooded areas over most of the northeastern part of the continent. The low evergreen plants sport bright red berries throughout the winter.

Wintergreen leaves are used to make a tea. (See Teas and Tisanes in the "Culinary Practice" chapter.) Pour two cups of boiling water over one tablespoon of dried leaves in a pot, and let it steep for a full five minutes. The fresh red "teaberries" are sweet and fleshy, with that same mild but distinctive wintergreen flavor. This flavor has long been popular in root beers and for chewing gum and candies. Though not actually a mint, wintergreen has a similar cooling effect in the mouth and is frequently used as a flavoring for the kind of refreshing candies generally called mints.

Be very careful to use only wintergreen *extract* or wintergreen *flavoring* for culinary purposes, and not wintergreen oil. The wintergreen oil on the market is actually intended as a liniment, and it is far too strong to be taken internally; it should not even contact the eyes or mucous membranes. Keep any products containing wintergreen oil out of the reach of children.

For an enjoyable variation, wintergreen extract or flavoring may be substituted for **vanilla** extract in cookies, coffee cakes, breakfast buns, or other baked goods. Use one-half teaspoon of wintergreen extract for each teaspoon of vanilla called for in your recipe.

woodruff/sweet woodruff/waldmeister (*Galium odoratum*): This sylvan herb, native to Europe and parts of Asia, is cultivated in gardens there and in temperate North America. Small white scented flowers bloom in spring and summer atop a stem that also bears one or two clusters of narrow, bright green leaves.

When fresh, woodruff leaves have no odor, but as they dry they develop a strong, long-lasting aroma, pleasantly reminiscent of new-mown hay and **vanilla**, which makes woodruff popular in potpourri and perfumery. Its best-known culinary use is to flavor German May wine—Rhine wines in which woodruff leaves have been steeped for a few hours and then strained out—and the punch called *Maibowle*, which is May wine plus sugar and strawberries, fortified with champagne and/or brandy; both of these beverages are traditionally drunk on May Day.

Woodruff supplies a slightly tart herbal lift to a type of German sausage. Custards and other milk dishes can be made more interesting by steeping a few leaves of woodruff in hot milk for a short time. The herb is sometimes used to scent hard candies, and also is used in the manufacture of several herbal liqueurs and digestives. May wine itself can be made into a jelly with unflavored gelatin.

Please note, however, that the pleasant vanilla scent of dried wood-ruff derives from coumarin, a substance toxic to the liver in large amounts. In the past, coumarin has been used as a flavoring, but it is now banned by the Food and Drug Administration as a food additive. Woodruff therefore should be used in the kitchen with caution, if at all.

Flavor Combinations

There are a number of standard flavor combinations, some of which have made a name for themselves and become essential to the cuisine of some part of the world. These traditional blends are listed here, along with a few less specific, but important, combinations.

Spices and herbs appearing in **bold** type are described in detail in the chapter on Individual Seasonings, where they are listed in alphabetical order.

HERB AND SPICE BLENDS

advieh: Like **curry powder**, this Iranian spice mix varies widely depending upon the dish it is intended to season. Advieh is not a hot blend; it typically includes **cinnamon**, **cardamom**, and **coriander** seed, while black **pepper**, **cloves**, **cumin**, **golpar**, and powdered **rose** petals may be added. The spices are ground and mixed.

If you do not wish to mix up an advieh for yourself, you might substitute **garam masala**, or simply **allspice**.

berbere: This fiery Ethiopian seasoning can be made either as a paste or a powder. It is used as a dip and is also an essential ingredient in many Ethiopian dishes. Most meat or fish stews call for a generous amount of berbere paste. Its primary ingredient is very hot **chile**, and indeed the word "berbere" is used to mean either the pungent sauce or the chiles themselves. This mixture is accented with a fair amount of **ginger**, **cloves**, and **cardamom**, and is further enriched by a selection of **ajowan**, **cinnamon**, **onions**, **garlic**, **coriander** seed, **fenugreek**, black **pepper**, or holy **basil**.

bouquet garni: A bouquet garni is widely used in classic French cooking, usually for flavoring stocks, soups, and stews. The standard ingredients are **parsley**, **thyme**, and **bay leaf**, although variations are allowed: if you think **rosemary** would taste good in the dish you are preparing, then add a sprig of it to your bouquet; if you favor the flavor of **celery**, then include a short stalk of it with the herbs. Leftover parsley stems may be used instead of the leaves. Tie a string around four to six stems of parsley (with or without the leaves), along with a sprig of thyme and one whole bay leaf. Dehydrated parsley flakes and dried thyme can be used if the seasonings are tied up in cheesecloth; use one teaspoon dried parsley and one-fourth teaspoon dried thyme to one bay leaf. In both cases, the string should be left long so that the herbs can be dangled in the liquid and then pulled out again at the end of the cooking. Discard the used bouquet.

chaat masala: This is a complex and exciting mixture of spices commonly used for snack foods in India. It is not added during cooking, but sprinkled lightly over the food afterward. Ingredients vary, but the blend frequently includes **salt, ajowan, amchur**, black salt, black **pepper, chiles, coriander** seed, dried pomegranate seeds, **asafetida**, dried **mint, ginger**, and/or **cumin** seed.

chili powder: In hopes of eliminating a lot of confusion, many food writers and food editors of the nation's newspapers are advocating the spelling "chile" for a hot pepper, while "chili" indicates the popular dish chili con carne, made with meat, chiles, spices, and other hotly debated ingredients. Chili powder is the spice mixture used to season this "bowl of red"; its essentials are ground red **chiles, garlic, cumin**, and strong-flavored Mexican **oregano**. Some manufacturers of chili powder include black **pepper, cloves, coriander** seed, or other seasonings.

Unfortunately, the confusion between "chile" and "chili" persists, and some recipes, often Indian recipes, say "chili powder" when they mean only ground-up red peppers. Since chili powder intended for chili con carne contains additional spices, it has a very different flavor from pure ground peppers and is less pungent.

crab boil/shrimp boil/shrimp spice: This flavorful combination of whole spices will guarantee that boiled crab, shrimp, crawfish, lobster, fish, and other seafood are never bland. The exact composition of the blend varies, but it generally includes **allspice, bay leaf, cloves, coriander** seeds, **fennel** seeds, **mustard** seeds, and black **pepper**. Sometimes **cinnamon, dill** seeds, **ginger**, or an herb such as **savory** may be added. Dried **chiles** or red pepper flakes allow you to adjust the piquancy to your own tolerance. Tie the whole spices up in a piece of cheesecloth, so they won't cling to the seafood.

Immerse the bag of spices in a large pot of cold water, allowing about one-fourth cup of mixed spices for one gallon of water. Bring the pot to a full rolling boil; let it boil for a good five minutes, to release the flavors of the spices, before adding the seafood. Don't overcook.

Crab boil does not need to be limited to seafood: vegetables such as new potatoes and corn on the cob are delicious when cooked with a small bag of crab boil spices in the cooking water. When cooking potatoes, keep checking the flavor; you might want to fish the cheesecloth bag out of the cooking liquid before the vegetables become too spicy.

curry powder: Traditionally in India, the spices used in a curry are selected individually, with attention given to the tastes of the eaters and to the main ingredients of the dish—a lamb curry, for example, gets a different combination of spices than a spinach curry—and authentic Indian recipes never call for curry powder.

To make a curry in the traditional Indian way, begin with whole spices. Toast them (see Toasting Spices in the "Culinary Practice" chapter), then grind them or, in some cases, you may leave them whole to be picked out or eaten as the diner prefers. Just as whole spices for curry are toasted, so a powdered spice must first be fried briefly in a little bit of hot ghee or vegetable oil, or lightly heated in a dry skillet; to a good Indian cook, a dish with "raw" spices is an embarrassment.

Clearly you do not want every curry dish to taste the same, but you can use a good, fresh commercial curry powder and still vary the flavor by adding a few extra spices suitable to each particular dish. Spices commonly found in commercial curry powders are **cumin**, **coriander** seeds, ground **chiles**, **turmeric** (the source of that yellow color), and **fenugreek** (for that distinctive "curried" smell). Other spices, such as **ginger**, **cardamom**, **cloves**, **nutmeg**, and **bay leaves**, are frequently included by various spice companies in their curry powders. You may consult your taste buds about adding **allspice**, **mace**, or **mustard** seed. To get the correct taste of a south Indian curry, you need a handful of **curry leaves**, if possible.

Curries are by no means confined to India. They have become an integral part of the cuisines of Thailand, Japan, and indeed all of the Orient, and of the Caribbean as well. Variations in the spicing and heat are made to suit the local palate.

Many ordinary American dishes profit from being "curried." A pinch of curry powder added to cooked egg yolks, along with mayonnaise, **salt**, and **pepper**, is a delicious flavor variation for deviled eggs, and a little curry powder mixed into the flour used for breading chicken pieces makes intriguing fried chicken. Sauté slices of tender young parsnips in butter with a sprinkling of curry powder. A teaspoonful of curry powder in a cup of sour cream or yogurt makes a terrific, tasty dressing for fruit salad. A pat of curry-seasoned butter melting over hot grilled meats and vegetables is exotic and exciting.

dhansak masala: *Dhansak* is a dish dear to the hearts of the Parsis in India. Made of a wealth of vegetables and pulses (lentils, split peas, and beans), it is served with a rice dish containing rich brown caramelized onions to make a complete Sunday meal. The dish is usually flavored with pieces of mutton or chicken and a special spice mixture called, appropriately enough, dhansak masala ("dhansak spices"). This mixture consists of a multitude of spices, including at least **cumin**, **coriander**, and **cardamom** seeds, black **pepper**, **cloves**, **cinnamon**, **turmeric**, and cayenne pepper or **chile** flakes. The spices are toasted in a dry skillet; be sure to start with the seeds, peppercorns, cloves, cinnamon, and so on, adding the turmeric and powdered or flaked chiles just at the end, so that the latter two spices will not burn. This masala mixture may also be used with any other lentil dish.

fines herbes: This mixture of finely chopped fresh herbs, popular in classic French cuisine, usually consists of **chervil**, **chives**, **parsley**, and **tarragon**. These four herbs can be used in equal parts, or you can make the mix a little heavier on the parsley and chives, depending on your own taste or what you have available. Finely chopped **marjoram**, **thyme**, **basil**, or **rosemary** are sometimes added to the mix.

The bright herbal flavor of this combination is especially good with cheese and egg dishes, *omelette aux fines herbes* being a prime example. It is also perfect for mushroom dishes. A dusting of powdered fines herbes enhances roast chicken and game hens; apply it just as the bird emerges from the oven. Don't be shy about using large quantities!

Since fresh chervil and tarragon are often difficult to find, you might try using dried herbs, in one-half the amount that you would use of fresh ones. Crush the dried leaves between your palms before adding them to the dish, or try refreshing them by soaking them for ten minutes in a tablespoon or so of white wine before using. You can also use Mexican mint **marigold** leaves as a substitute for tarragon.

five-spice powder: This ancient Chinese spice blend is traditionally made of equal quantities of intensely aromatic spices: finely ground **Szechwan peppercorns**, **star anise**, **cloves**, cassia **cinnamon**, and **fennel** seed. Variations sometimes use **anise** seed instead of star anise and cinnamon for cassia; the Szechwan peppercorns may be replaced with red pepper or **ginger**. Sometimes the number *five* is taken symbolically (for health) rather than literally, and the mixture may use any or all of the above spices, plus perhaps a little ground **licorice** root.

Five-spice is used with "red-cooked" meats and poultry (that is, braised over low heat in soy sauce). The powder should be used sparingly, by the half-teaspoonful. A vegetarian version of red-cooked meats is made by braising cubes of tofu in soy sauce and five-spice powder; the wonderfully complex flavor of the spice mixture wakes up the bland bean curd and makes a very interesting dish.

This seasoning is sometimes combined with twice the quantity of fine **salt** to make a seasoned salt.

Five-spice powder easily crosses over into traditional American cuisine and works wonders with spice cakes and cookies, in pumpkin pie, or lightly sprinkled on baked squash. Add a little to doughnuts or to the batter for fried chicken.

garam masala: *Garam masala* is often said to mean "warm spices," but a more accurate description is "warming spices." This mixture is not hot on the tongue; it is mellow rather than pungent, but it contains many of the spices that increase internal body heat, and for this reason Indian cooks use it much more in the wintertime than in the summer.

The combination, traditionally from northern India, is popular all over the subcontinent. Formulas vary, but typically a garam masala mixture consists of the aromatic spices **cardamom** seeds, **cinnamon**, **cloves**, black **pepper**, **cumin**, and **nutmeg** and/or **mace**. **Coriander** seed is often added as well. After toasting gently, the spices may be used whole or ground to a powder. Add whole spices early in the cooking process, but the powder should go in toward the end of the preparation.

This aromatic blend has myriad uses. It goes into a sauce for meat and poultry dishes, it is good with vegetables and lentils; many savory rice dishes are spiced up with this sweet mixture. And garam masala happily goes cross-cultural: the powder can be sprinkled as a garnish on cheese and soufflés or other egg dishes from almost any cuisine. It is delicious in cream sauces or soups; in these dishes, to keep the dark powder from spotting the cream, it is advisable to simmer the whole spices in the sauce and strain them out at the end. Put just one-eighth teaspoon of garam masala in a pot of breakfast oatmeal or in a sweet-potato casserole; the spice mixture will make a world of difference to the dish without calling attention away from the main ingredient to itself. Use garam masala instead of the usual pumpkin pie spices or to make an interesting variation on Dutch *speculaas* cookies.

gremolata: Gremolata is a sprightly mixture of chopped flat-leafed **parsley**, grated **lemon** zest, and fresh **garlic**, the latter very finely chopped. These ingredients may be used in roughly equal parts, or the garlic may be reduced and/or the parsley increased, according to the dish and your own taste. Sometimes the mixture is moistened with a little good-quality olive oil. Gremolata adds interest to potato and other vegetable dishes that might otherwise tend to be bland. Traditionally, it provides an important finish to *osso bucco* and other dishes of Italy and southern France. Try sprinkling it over a roast chicken, still hot from the oven. A separate bowl of this mixture may also be passed at the table.

This same flavor harmony, expressed in an Oriental key, is found in the traditional Middle Eastern garnish made of **parsley**, onion slices, and **sumac**. Like gremolata, this garnish uses flat-leaf parsley; the onions are akin to garlic but, since they are sweeter and milder, a larger proportion of onion can be used; and sumac has a fruity sourness that corresponds to **lemon**. This Middle Eastern *piyaz* is excellent with grilled meats, especially with ground-meat kebabs, which are molded on a skewer and grilled.

herbes de Provence: This mixture is so delightfully perfumed that at least one chef cannot leave it in the kitchen; she ties up a tablespoon of it in cheesecloth and attaches it under the tap as she runs the water for a soothing, scented bath. More traditionally, it is sprinkled over grilled meats or roast chicken, or incorporated into a bread stuffing. Herbes de Provence is renowned with roast lamb. It is excellent in tomato sauces and as a coating for cheese, so naturally it's great on pizza.

If you like **lavender** ice cream, you might try making herbes de Provence ice cream in just the same way. Or grind and sift your favorite mixture, and incorporate the little bits into the frozen custard. Your guests will be baffled about the identity of its complex, invigorating flavor. Garnish with a sprig of any one of the component *herbes*.

Just as ground **sage** is incorporated into the batter for biscuits to serve with ham or chicken, a little ground-up herbes de Provence will make excellent biscuits to accompany roast lamb.

Recipes for this favorite mix vary widely. Adjust the following suggested recipe according to what you like and what you have available: one part **rosemary**, one part **lavender**, two parts **basil**, two parts **thyme**, four parts summer **savory** (or three parts winter savory), six parts **tarragon**, six parts **marjoram**. Some people also add one part **fennel** seed. The herbs can be fresh or dried. If possible, use whole leaves, then grind up a small amount of herbes de Provence when you need it.

jerk seasoning: "Jerk" or "jerked" pork, beef, and chicken are the latest versions of a centuries-old method of preserving meat by cutting it into thin strips and sun-drying it or smoking it over green wood. In North America, the resulting product is usually called jerky, a name that comes to us via Spanish from Quechua, the language of the Inca civilization. Often, various seasonings are rubbed into the meat while it is still fresh, to preserve and flavor it. Those seasonings from the Caribbean, especially Jamaica, are particularly tasty, and lately this version of jerk meat has become popular. It can be served hot as a main dish, or cold slices may top a salad.

Today, jerk seasonings are often treated simply as a scrumptious, spicy-hot marinade for grilled meats; if you want a marinade rather than a dry rub, just add to these seasonings some flavorful acid—such as vinegar, **lime** or **orange** juice, or **tamarind** pulp—until the mixture is sufficiently liquid. Adjust the heat of your seasonings to your own taste by altering the amount of hot red pepper; the proportions given here are quite incendiary.

Cut the meat into thin strips, about one-fourth to one-half inch thick, and rub with lime juice. (See Citrus in the "Culinary Practice" chapter for advice on juicing limes.) Mix into a paste: one tablespoon ground **allspice**, one teaspoon dried red pepper flakes or minced fresh hot pepper of your favorite type, one-half teaspoon freshly grated **nutmeg**, one-half teaspoon black **pepper**, one-half teaspoon **thyme**, one clove **garlic**, one scallion with top, one-half small white onion, and one-half teaspoon vegetable oil. Rub this paste onto meat strips and let marinate for at least an hour; then grill or broil the meat. This recipe is best for beef. For pork, add one-half teaspoon of **sage**; for chicken, add one teaspoon of **marjoram** and substitute two more scallions for the white onion.

khmeli-suneli: A variety of spice mixtures are sold under this name in the markets of Georgia, the ex-Soviet republic located between the Black Sea and the Caspian. These elaborate combinations generally feature ground **coriander** seed with several dried herbs such as **basil**, summer **savory**, **marigold** petals (either *Calendula* or *Tagetes* marigolds), **mint**, **fenugreek** leaves, and **dillweed**. They may also contain ground **fenugreek** seeds, **cloves**, **cinnamon**, black **pepper**, **celery** seed, or **paprika**.

Used in meat and vegetable soups or stews, and also in meat marinades and bean dishes, khmeli-suneli is similar to **curry powder** in that the choice of ingredients and their proportions vary with the dish to be seasoned.

mirepoix: This is another important flavor combination from classic French cuisine, consisting of finely diced carrots, celery, and onions, and occasionally other aromatic vegetables such as leeks. Sometimes ham or bacon is included, in which case a **bay leaf** should be added. These aromatic vegetables are used in making rich sauces or in braising meats; generally they are simmered for a long time until all their flavor is given up to the liquid. Afterward, they are strained out, pressed against the sieve to extract all their juices, and then discarded. The trick in preparing mirepoix is to dice everything the same size so it all cooks in the same length of time.

mixed spice/pudding spice: In some older cookbooks, especially books from England or Australia, you may find a recipe calling simply for "mixed spice." Several mixtures used to be sold under this name; some were also known as "French spice" or "spice parisienne."

The spices selected for these blends are the sweet, warm favorites of English cooking, of the sort that make up the pumpkin pie spice blends found in the supermarket today. A representative combination might include **allspice**, **cinnamon**, **clove**, **coriander**, **ginger**, and **nutmeg**. In England, mixed spice blends flavor spice cakes, fruit cake, mincemeat, and other sweets, and thus are also known as pudding (dessert) spice. Try adding a teaspoonful to your favorite pecan pie recipe. In France, this blend is most often used with game or beef, but is also used in almost any other meat dish.

In a similar way, all-purpose blends of dried mixed herbs were sometimes put together by local grocers or spice companies for the convenience of the home cook. These often included **marjoram**, **sage**, **savory**, and **thyme** with a little black **pepper**, and bore a resemblance to today's ready-made poultry seasoning.

If you encounter an old recipe calling for mixed herbs, your own personal seasoning savvy—and, perhaps, this book—will guide you in choosing appropriate herbs for your dish.

mulling spices: Beverages such as cider, wine, ale, or beer that
have been heated, sweetened, and flavored with spices are said to be
"mulled." (The word is of uncertain origin.) Whole spices are best
for this process, to keep the liquid clear, but you may tie up powdered spices in a little muslin bag and steep them in the hot liquid.
Warm, sweet, aromatic spices are preferred for mulling, such as
those in the following typical recipe.

To about three cups of liquid, add four whole **cloves**, one stick of
cinnamon or six **cassia** buds, four blades of **mace** or one-quarter of
a whole **nutmeg**, four **cardamom** pods, and seven **allspice** berries.
Add sugar to taste and three or four very thin **orange** slices. Simmer—never boil—for twenty minutes or longer. Strain and serve
hot on a chilly evening.

panch phoran: The name of this Bengali seasoning means "five
seeds," and the five referred to are **cumin, fennel, fenugreek, mustard** (brown, yellow, or a combination of the two), and **nigella**. The
seeds are used whole, and combined in equal proportions.

The mixture is aromatic and strong-flavored, and is used to give a
kick to lentil dishes and to vegetables such as eggplant, squashes,
and tomatoes.

Panch phoran needs to be briefly cooked, either toasted in a dry
skillet or fried momentarily in some kind of fat. Be sure to use
medium to medium-high heat—heat too high will burn the seeds
and ruin their flavor—and stop immediately when the mustard
seeds just begin to pop. In Bengal, this step is followed by the quick
addition of the vegetables and any other ingredients, which cools
the pan.

This unique seasoning should not be confined to the east of India.
If you like its strong, bitter, complicated taste, experiment with
using it in any vegetable dish. A mundane bowl of boiled cabbage
becomes exciting with a sprinkling of toasted panch phoran; a bowl
of frozen peas turns exotic when topped with a drizzle of clarified
butter or ghee and lightly fried seeds. Even better, pour this panch
phoran butter over popcorn!

persillade: This popular French seasoning is composed of finely
chopped **parsley** and **garlic** or shallot, also finely chopped. Proportions are about one clove of garlic or shallot to one-fourth cup

chopped parsley. Persillade is added to dishes just at the end of the cooking time. It is good mixed with the juices of the dish; sometimes it is mixed with bread crumbs and melted butter, or it may be sautéed in olive oil. If you like, you may add other Francophile herbs, such as **chervil, chives, savory, thyme,** or **tarragon**. This makes a bright, fresh topping for poultry, seafood, vegetables, or omelettes.

pickling spice: For pickling, it is best to use whole spices rather than ground ones, in order to keep the liquid light and clear. The spices can be tied up in a piece of cheesecloth, then removed before the pickles are bottled; or, if you prefer, the spices can be included in the jar with the pickled vegetables. Be sure to give the whole spices enough time in the hot vinegar or other pickling liquid to release their flavors; ten minutes will give a delicate seasoning, or an hour of gentle boiling will extract their full taste.

Pickling spices generally include **mustard** seeds, **celery** seeds, and **allspice** berries, with the additional choice of **turmeric, cinnamon** bark or **cassia** buds, whole **cloves, ginger** root, or blade **mace**; sometimes **coriander** seeds and **bay leaves** are also added. Adjust the piquancy with small dried **chiles**. Be sure to use turmeric in mustard pickles. Dill pickles are flavored with both **dill** seeds and dillweed, the ideal form being sprigs of fresh dill with the seed heads still attached. And there is no pickle that is not improved by a few blades of mace.

Soak the mustard seeds in *cold* water for a few minutes before using, and they will have a much nicer flavor.

quatre épices: There are many different recipes for this popular French spice blend, and despite the name they are not all restricted to only four ingredients! A common quartet is white **pepper, cinnamon, nutmeg,** and **clove**, with the pepper predominant, accounting for about half the total spice in the mixture. Black pepper may be used instead of white. Cloves are added with caution—for one-fourth cup of the blend, only about a teaspoon of cloves would be used—to avoid overpowering the other spices. **Ginger** is sometimes substituted for the cinnamon, or simply added to the mix. This finely ground blend is commonly used to season patés, as well as meats and sausages.

In France, **allspice** is often called quatre épices because the flavor of this spice resembles a combination of pepper, cinnamon, nutmeg, and clove. The name is also occasionally applied to **nigella**, for no good reason whatsoever.

When Tunisian cooks make their version of quatre épices, they include **paprika** and a few dried **rose** buds.

ras el hanout: This is the most famous spice blend in Morocco, a country well versed in spices. The Arabic name means "head of the shop," which implies something like "top of the line." Each spice merchant boasts about the number of ingredients in the company's own special version of this mixture—as many as forty-five have been claimed—but keeps the recipe a secret.

Among the typical spices in this blend are **allspice, cardamom, cinnamon, clove, galangal, ginger, grains of Paradise, mace, nutmeg**, black **pepper, rose** buds, and **turmeric**. And those are just the basics!

Ras el hanout is used especially for game, but also seasons tajines, rice, and many other dishes, especially in winter. It is regarded as warming and as an aphrodisiac.

Use ras el hanout in lamb, beef, or venison stews. Mix a rounded teaspoonful of this blend with a pound of ground beef for the best burgers ever.

sambar powder/sambhur masala: In the south of India, vegetarian eaters consume a tangy lentil and vegetable soup called *sambar* virtually every day. Sambars get their tang from **tamarind**, and their distinctive flavor from a complicated blend of seasonings known as sambar powder. The blend is available ready-made at Indian shops, or you can put it together at home from scratch.

An unusual feature of sambar powder is the inclusion of *dals* (dried beans, peas, or lentils), in keeping with the Indian cook's use of these legumes to vary the flavor and texture of a dish. Sambar powder always contains *toovar dal*, or *toor dal* (very like yellow split peas), and often other dals as well; these are available at Indian grocers. Standard spices in the mixture are **asafetida**, red **chiles**, **coriander** seed, **cumin, fenugreek** seed, black **pepper**, and **turmeric**. On occasion, **cinnamon, curry leaves, mustard** seed, or

sesame are added. The dal is fried and the spices are toasted; then all are coarsely ground together to make the sambar powder.

shichimi togarashi/shichimi: In Japan, this exciting blend of "seven flavors" (the literal meaning of the name) seasons soups, noodles, and *yakitori*—skewered morsels of grilled chicken, usual-ly served with drinks. The mixture of coarsely ground dried spices is generally used as a table condiment, sprinkled over finished dishes according to taste.

Based on red **chile** powder, shichimi is hot enough to singe the top of the tongue and to start a tingle in the lips, but the six other flavors do come through as well. These ingredients vary slightly with the source of the recipe, but usually include sansho (see **Szech-wan peppercorn**), black and white **sesame** seeds, dried citrus peel, **poppy** seed, hemp seed, and seaweed. Sometimes **perilla,** flaxseed, rape seed, or **ginger** are selected in addition or instead.

za'atar: This herb and spice combination is a favorite in the Middle East. Bread dipped into za'atar and olive oil is a great appetizer, side dish, or snack. Flat pita breads baked with a za'atar topping are especially popular for breakfast.

The basic ingredients of za'atar are ground **thyme**, ground **su-mac**, toasted **sesame seeds**, and a little **salt**. The ratios vary, de-pending on whose recipe you are using, but four parts herb to two parts sumac to one part sesame seed is a good place to start. Salt to taste, and adjust the other ratios to your own desire. Regional varia-tions from Sudan to Lebanon may include additional seasonings ranging from **dill** seed to **lemon** zest to ground pistachios. Often, thyme is replaced by some other dark-flavored green herb growing locally, such as wild thyme (another *Thymus* species), **hyssop**, **mar-joram**, **oregano**, or **savory**.

COMPOUND BUTTERS

Softened, *not* melted, butter can be mixed with seasonings of all kinds, molded into the desired shapes and chilled in the refrigerator to harden. This is a lovely way to introduce another flavor, and if extra pats of butter are served at the table, the diners can help themselves to more until they're satisfied. Use an unsalted butter and add the appropriate amount of **salt** when you're mixing up the compound butter.

Fresh green leaves, very finely chopped, are a favorite choice for these butters. Several different herbs can be mixed if you choose. **Chives**, **coriander** leaf (cilantro), **lemon verbena**, **parsley**, and **tarragon** leaves are often used. Use about one tablespoon of the herb for one-fourth cup of butter (half a stick), except for the stronger-flavored chives; try half as much of the chives. Of course, all these measurements can be changed according to your own taste.

The herb butters are particularly useful to serve with roasted or grilled meats, fish, and poultry; all of these respond well to herbal flavors. *Maître d'hôtel* butter, from classic French cuisine, is a favorite for this purpose. It combines chopped parsley and **lemon** juice with butter, salt and white **pepper**. Proportions vary according to taste; usually more lemon is used for fish, and a little finely chopped shallot may be added for meat. Slather on the butter before, during, and after roasting, but save it to season grilled meats on the platter, as the grilling process is generally at temperatures too hot for the delicate leaves to withstand.

Compound butters can be frozen for later use.

FLAVORED VINEGARS

Different vinegars have their own very pronounced flavors, and this must be considered when choosing one for use in a given recipe. Vinegar is created by the action of certain bacteria on alcoholic liquids such as cider, wine, and fermented fruits or grains; these bacteria convert alcohol to acetic acid and water, retaining more or less of the flavor characteristics of the original liquid.

The filmy "mother of vinegar" or "vinegar plant" sometimes found floating in vinegars is composed of these acetifying bacteria; the "mother" is not harmful but it is certainly unattractive. Being composed of numerous minuscule creatures, it is impossible to filter out. Its growth is encouraged by temperatures between about 60°F and 85°F, and is discouraged by higher temperatures. If your vinegar is cloudy, with a spooky vinegar mother floating in it, your only hope is to pour off as much as possible of the gelatinous mass (saving a little for a future vinegar starter) and boil the rest of the liquid to stop the fermentation; then cool it immediately and filter it into bottles that have been sterilized in boiling water.

Don't use aluminum pots or utensils when dealing with vinegar. Vinegar should be stored, tightly sealed with a nonmetallic lid, in a cool, dark place.

Although vinegar can be made from virtually any fermentable substance—fruits, grains, honey, molasses, milk, coconut, and so forth—you probably will be dealing with one of the basic, common types listed below. Most vinegars on the market have a 5 percent acidity; wine vinegars usually start at 6 percent; Oriental rice vinegars are only about 4 percent acidity. Check the label. Sometimes acidity is indicated in "grains," with fifty grains corresponding to 5 percent acid. Higher acid content usually indicates a sharper taste, but other factors in its manufacture can make an acidic vinegar more mellow.

When selecting vinegars it is important to keep in mind that a minimum of 5 percent acidity is needed for safe pickling and preserving, which eliminates from consideration rice vinegar, most fruit vinegars (both described below), and homemade vinegars of indeterminate acidity. The only absolutely foolproof method for choosing a suitable vinegar for your recipe is to use your own taste buds, but the discussion of common vinegar types that follows should be helpful. If you find a vinegar too sharp for the salad dressing or fruit pie or whatever dish you're making, dilute it with a little water, wine, or cider, whichever is appropriate: about one tablespoon per cup of vinegar will make a difference. Do not, however, dilute vinegar used for pickling or preserving.

The Common Types of Vinegar

Cider Vinegar

Made from apple cider, this is the most popular vinegar in the United States. It has a pale straw color and a mild fruity flavor that makes it usable in almost any recipe. It is a good choice for marinating chicken or turkey. Sprinkle a teaspoon or two of good cider vinegar over apple pie filling to keep the pie from being cloyingly sweet.

Distilled Vinegar/White Vinegar

Although it is sometimes called "white," distilled vinegar is as colorless as water. This type of vinegar is preferred for pickling because it has very little flavor of its own and, being distilled, will not develop a mother plant. It is generally the cheapest kind of vinegar. Sometimes it consists of pure acetic acid, diluted to a 5 percent acidity.

Wine Vinegars

Wine vinegars, made from red or white grapes, are the only option for the French chef. They should also be used for all Italian and Spanish recipes. The taste generally has a pleasant grapey mellowness, but the quality can vary according to the grade of wine begun with. Champagne vinegar is light and mild; sherry vinegar is full-flavored but still not sharp. White-wine vinegar is preferred for vinaigrette, white sauces, and most marinades. It is also the usual choice for making flavored vinegars (see below). Red-wine vinegar is used in cooking red cabbage and other dark or strong-flavored vegetables such as onions to enhance their flavor as well as to reduce their odor. It is often used instead of, or along with, red wine in marinating beef or venison.

Wine vinegars are often 6 percent acetic acid, and can be higher; on the other hand, they are sometimes diluted with water to 5 percent acidity.

Malt Vinegar

Made from malted barley or other grains, and tinted with a little caramelized sugar, this pungent, sweet vinegar is *the* condiment for English fish and chips. Malt vinegar also gives authentic flavor to chutneys. It is good in marinades for dark, fatty fish such as mackerel. Choose malt vinegar when making any traditional British or Australian dish.

Balsamic Vinegar

Made from sweet white grapes, this dark, thick, strong vinegar is properly aged for a minimum of twelve years in a series of ever-smaller wooden barrels that give it its color and flavor. A specialty of Modena and its surrounding region in northern Italy, balsamic vinegar has recently become popular in the United States, bringing with it a lot of confusion among buyers as numerous vinegar makers jump on the bandwagon, manufacturing their own versions with sweeteners, colorings, and flavorings. As a result, there are many different qualities of balsamic vinegar on the market, and not even price is a reliable guide in making a choice. One infallible way to choose a good balsamic vinegar, however, is to look for certification from the Italian government that this is genuine *aceto balsamico tradizionale di Modena*—the wording must be exact. However, these imported, certified balsamic vinegars can be wildly expensive. Another method is to purchase a small amount of any dark, viscous balsamic vinegar in your price range and test it in your kitchen, repeating until you find one that meets your needs and standards of taste.

Intensely flavored balsamic vinegar should be used as a seasoning rather than an ingredient, with just a few drops sprinkled on any course, from fish to polenta to salad to dessert. Do not cook with balsamic vinegar. Combined with a good extra-virgin olive oil, it makes a wonderful vinaigrette for a plain green salad!

Rice Vinegar

Oriental vinegars are often derived from rice; they are very mild, usually diluted to 4 percent acidity. Light, clear rice wine vinegar is important in Japanese cooking; the seasoned variety—with sugar, **salt** or soy sauce, and sometimes MSG, added—is essential for sushi, and it is the vinegar used in making pickled **ginger**. *Omeboshi* vinegar is what remains from the process of making Japanese salted plums; it is fruity, low in acidity, and colored ruby red with red **perilla** leaves. The Chinese make white, red, and black vinegars from rice mash; the white is used for sweet and sour dishes and for sauces and salads. Black vinegar is sweetened and flavored, reducing its acidity to very low levels; it is popular in Pacific Rim cookery.

Fruit Vinegars

"Fruit vinegar" sometimes refers to vinegars made from fermented fruits, and sometimes indicates a vinegar, usually white-wine vinegar, which has been flavored by steeping ripe fruits in it. Either way, these vinegars are delicious, often low in acid and mellow, with rich fruit flavor. Use them to marinate meats or, sparingly, to flavor fruit desserts.

How to Make Flavored Vinegars

Flavored vinegars are delicious and versatile. They can be used as major ingredients in salad dressings or marinades, or a couple of tablespoons can be added as a flavoring to sauces, stir-fries, fruit dishes, or even stuffing for poultry. The best vinegar to use is white wine vinegar; champagne vinegar is light and mild, and will not overwhelm delicate flavors. The procedure is simple: the flavoring materials—herbs, spices, fruits, or flowers—are loosely packed in a clean *glass* bottle; then hot, but *not boiling*, vinegar is poured over them. When cool, the bottle should be sealed tightly, preferably with a cork. A plastic lid is acceptable, but do not use a metal cap. Set the bottle aside for about two weeks in a cool, dark place. Shake it occasionally. After the vinegar has absorbed the taste of the introduced flavoring to your satisfaction, strain it and pour it into new bottles that have been sterilized in boiling water. It is sometimes necessary to filter the vinegar at this stage, using a coffee filter or several layers of cheesecloth, to remove tiny particles of the flavoring material. At this point, for an attractive presentation, you can insert a fresh sprig of herb, including its flower when possible, or some whole spices. Note, however, that this will gradually increase the intensity of the flavor; for this reason, it is not recommended that you add whole **garlic** cloves to your finished garlic vinegar. If you desire to make that popular Oriental condiment, pickled garlic, see garlic in the "Individual Seasonings" chapter.

Herbs of all kinds are popular flavorings for vinegar. **Tarragon** is a great favorite, as are **dillweed** and **basil**; opal basil and the other purple basils will turn the vinegar a lovely pink color. Fresh herbs are far superior but dried herbs can be used; much less dried herb is needed—one-third to one-half the amount of fresh herb is sufficient.

Chiles also can color the liquid. Many delicious hot sauces consist simply of chiles stuffed into a bottle of vinegar. This is a common condiment in the South for boiled spinach and other greens.

Rose petals give up a delicate color and aroma to a champagne or other very mild vinegar. Of course, you must be sure that the flowers have been grown organically. The rose fragrance can be enhanced by a little rose water, added when bottling the finished vinegar—only one-fourth teaspoon rosewater will enchant an entire pint of vinegar.

Not only leafy herbs, but whole spices and seeds are excellent for flavoring vinegars. **Cinnamon** sticks, whole **allspice** berries or **cloves,** and blade **mace** make particularly interesting vinegars for use in marinades and in savory sauces, as do **mustard** seeds, **celery** seeds, and **fennel** seeds. Add one inch of cinnamon stick, or approximately one-half teaspoon of whole seeds or spices, to a cup of hot vinegar and let it steep for two weeks; then strain the liquid into a clean bottle with a tight lid or stopper. A good-quality red wine vinegar flavored with spices is useful in marinades, salad dressings, and sauces.

Fruits are a very popular flavoring for vinegars. Berries work particularly well, but citrus, tropical fruits, pears, and peaches all have something interesting to add. Choose your vinegar to complement the taste of the fruit, and proceed in the manner described above. All the fruits should be very ripe, but not necessarily fresh: frozen berries work very well.

FLAVORED OILS

These popular seasonings are a great way to add flavor to a dish. Commercial flavored oils vary widely in quality and intensity, and should be sampled before using, if not before purchasing. Usually they are quite strong in flavor because they are prepared with the pure essential oils instead of the whole spice or herb itself. Generally, these infused oils are best used only by the spoonful, as a flavoring.

The essences that flavor commercial oils have been sterilized; if you make your own infused oils, you need to be aware of the danger of potentially fatal poisoning. Homemade flavored oils, while delicious, must be handled carefully to avoid running the risk of botulism. This serious (often fatal) type of food poisoning is caused by toxins produced by various types of *Clostridium botulinum* bacteria, which commonly live in the soil and grow readily in the absence of oxygen and acids; thus botulism is often associated with improper canning of foods, and is also a danger in storing herbs and low-acid vegetables under oil. Low temperatures will retard the growth of these bacteria and their production of toxins, but the ordinary family refrigerator, especially if frequently opened, may not be cold enough to inhibit all types. The toxins are generally not detectable by taste or smell! While the spores of these bacteria are resistant to heat, the lethal toxins themselves can be destroyed by mere boiling for a few minutes. So if you make your own flavored oil with herbs or garlic or other low-acid plant material, be sure to clean the herbs well, keep the oil in the coldest part of the refrigerator, and use it up within a short time. Whenever possible, heat the oil before using it. If you are suspicious of its safety, don't hesitate to throw it out!

Besides being flavored with spices and herbs, oils have their own natural flavors. **Mustard**, olive, **sesame**, **coconut**, and the nut oils are all strongly flavored, while corn oil and peanut oil are of moderate intensity, and most of the seed oils—canola, safflower, and sunflower seed oil—are mild and neutral in taste.

Index

Note: Seasonings in **bold** are listed in alphabetical order in the "Individual Seasonings" chapter. Don't insist on orthographical precision when looking for a name, because variations in spelling abound and they cannot all be listed here.

coriander, 99; *also* 14, 16, 57, 80,
        96, 97, 104, 107, 109, 114,
        116, 146, 149, 156, 161, 177,
        179, 182, 235, 236, 237, 238,
        242, 243, 235, 246, 248
Coriandrum sativum - **coriander**, 99
coumarin, 228, 233. *See also* **vanilla**;
        woodruff
crab boil, 236
Crocus sativus - **saffron**, 194
cubanelle, 71. *See also* **chiles**
cubeb pepper, 181. *See also* **pepper**
culentro, 103. *See also* **coriander**
cumin, 103; *also* 4, 14, 34, 41, 56,
        57, 86, 97, 101, 111, 114,
        115, 149, 170, 222, 235, 236,
        237, 238, 240, 244, 246
Cuminum cyminum - **cumin**, 103
Curcuma longa - **turmeric**, 226
curly parsley, 179. *See also* **parsley**
curry leaf, 105; *also* 62, 97, 100,
        200, 238, 246
curry powder, 237
Cymbopogon citratus - **lemon grass**,
        141

datil pepper, 69. *See also* **chiles**
daun kesom, 103. *See also*
        coriander
daun pandan - **screwpine**, 209
daun salam - **salam leaf**, 200
Devil's dung - **asafetida**, 45
dhana-jeera, 101. *See also*
        coriander
dhansak masala, 238
Dianthus caryophyllus - clove pink,
        91. *See also* **clove**
dill, 105; *also* 8, 14, 18, 20, 30, 32,
        58, 64, 111, 113, 149, 161,
        177, 236, 242, 245, 247, 252
dillseed, 108. *See also* **dill**
dillweed, 105. *See also* **dill**
dried lime, 149. *See also* **lime**

Elettaria cardamomum - **cardamom**,
        58
epazote, 109; *also* 110, 177
Eryngium foetidum, 103. *See also*
        coriander

fagara - **Szechwan peppercorn**, 219
fennel, 109; *also* 4, 14, 41, 56, 57,
        104, 132, 144, 170, 177, 236,
        239, 241, 244, 253
fennel flower, 170. *See also* **nigella
        seed**
fenugreek, 113; *also* 107, 148, 149,
        235, 237, 242, 244, 246
filé - **sassafras**, 205
fines herbes, 238
five-spice powder, 239; *also* 112
flat-leafed parsley, 180. *See also*
        parsley
flavoring, 1
flax seed, 1, 14, 247
Foeniculum vulgare - **fennel**, 109
folia malabathri, 61. *See also* **cassia**
food-of-the-gods - **asafetida**, 45
fragrant lime leaf, 150. *See also* **lime**
French marigold, 156. *See also*
        marigold
French spices, 243

galanga - **galangal**, 116
galangal, 116; *also* 5, 118, 246
galingale, 119. *See also* **galangal**
Galium odoratum - **woodruff**, 232
garam masala, 239
garden angelica - **angelica**, 39
garden basil - **basil**, 48
garden sage - **sage**, 198
garden thyme - **thyme**, 224
garlic, 119; *also* 2, 11, 15, 42, 45,
        49, 80, 91, 92, 101, 102, 114,
        124, 132, 135, 160, 161, 163,
        170, 180, 193, 199, 214, 224,
        235, 236, 240, 241, 252, 254
garlic chive, 73. *See also* **chive**

kencur, 117. *See also* **galangal**
kentjoer, 117. *See also* **galangal**
kewda water, 211. *See also*
    screwpine
kewra water, 211. *See also*
    screwpine
key lime, 148. *See also* **lime**
kha, 116. *See also* **galangal**
khmeli-suneli, 242
khulinjan, 116. *See also* **galangal**
khus khus - **poppy seed**, 187
kinome, 220. *See also* **Szechwan**
    peppercorn
kitchen thyme - **thyme**, 224
knotted marjoram - **marjoram**, 156
kosher salt, 204. *See also* **salt**
krachai, 117. *See also* **galangal**
kumquat, 26

la dua - **screwpine**, 209
laksa leaf, 103. *See also* **coriander**
langkuas, 116. *See also* **galangal**
laos, 116. *See also* **galangal**
large cardamom, 61. *See also*
    cardamom
laurel leaf - **bay leaf**, 52
Laurus nobilis - **bay leaf**, 52
Lavandula angustifolia - **lavender**,
    136
lavender, 136; *also* 4, 200, 241
leavenings, 20
lemon, 139; *also* 20, 25, 26, 32, 39,
    49, 50, 64, 75, 107, 108, 114,
    139, 142, 143, 144, 145, 146,
    147, 148, 159, 160, 161, 162,
    163, 173, 188, 190, 193, 198,
    202, 214, 215, 219, 222, 223,
    224, 240, 247, 248
lemon balm - **balm**, 47
lemon basil - **basil**, 51
lemon geranium, 208. *See also*
    scented geranium
lemon grass, 141; *also* 5, 96, 116,
    39, 143, 224

lemon salt, 30
lemon thyme, 224. *See also* **thyme**
lemon verbena, 143; *also* 48, 139,
    150, 224, 248
lenguas, 116. *See also* **galangal**
lesser galangal, 117. *See also*
    galangal
lesser ginger, 117. *See also* **galangal**
licorice, 144; *also* 32, 41, 128, 151,
    239
licorice basil, 51. *See also* **basil**
lime, 146; *also* 26, 96, 97, 115, 116,
    125, 143, 182, 204, 223, 242
limou amani, 149. *See also* **lime**
limoun Basra, 149. *See also* **lime**
linden, 152. *See also* **lime**
lin seed - flax seed, 1, 14, 247
lipstick tree - **annatto**, 43
liquorice - **licorice**, 144
long pepper, 183. *See also* **pepper**
loomi, 149. *See also* **lime**

mace, 152; *also* 92, 172, 173, 238,
    240, 244, 245, 246, 253
magrut, 150. *See also* **lime**
mahaleb, 153; *also* 158
mahleb - **mahaleb**, 153
malabathrum, 61. *See also* **cassia**
malagueta pepper - **grain of**
    Paradise, 129. *See also*
    chiles
malepi - **mahaleb**, 153
mango powder - **amchur**, 38
marigold, 154; *also* 198, 242
marjoram, 156; *also* 16, 50, 128,
    176, 177, 207, 238, 241, 242,
    243, 247
masterwort, 39. *See also* **angelica**
mastic, 157
meadow saffron, 195. *See also*
    saffron
melegueta pepper - **grain of**
    Paradise, 129
melissa - **balm**, 47
Melissa officinalis - **balm**, 47

Order Your Own Copy of
This Important Book for Your Personal Library!

SEASONING SAVVY
How to Cook with Herbs, Spices, and Other Seasonings

_____ in hardbound at $39.95 (ISBN: 1-56022-031-7)

_____ in softbound at $24.95 (ISBN: 1-56022-032-5)

COST OF BOOKS_____

OUTSIDE USA/CANADA/
MEXICO: ADD 20%_____

POSTAGE & HANDLING_____
*(US: $3.00 for first book & $1.25
for each additional book)
Outside US: $4.75 for first book
& $1.75 for each additional book)*

SUBTOTAL_____

IN CANADA: ADD 7% GST_____

STATE TAX_____
*(NY, OH & MN residents, please
add appropriate local sales tax)*

FINAL TOTAL_____
*(If paying in Canadian funds,
convert using the current
exchange rate. UNESCO
coupons welcome.)*

☐ **BILL ME LATER:** ($5 service charge will be added)
(Bill-me option is good on US/Canada/Mexico orders only;
not good to jobbers, wholesalers, or subscription agencies.)

☐ Check here if billing address is different from
shipping address and attach purchase order and
billing address information.

Signature_____

☐ **PAYMENT ENCLOSED: $**_____

☐ **PLEASE CHARGE TO MY CREDIT CARD.**

☐ Visa   ☐ MasterCard   ☐ AmEx   ☐ Discover
☐ Diners Club
Account # _____

Exp. Date _____

Signature _____

Prices in US dollars and subject to change without notice.

NAME _____

INSTITUTION _____

ADDRESS _____

CITY _____

STATE/ZIP _____

COUNTRY _____ COUNTY (NY residents only) _____

TEL _____ FAX _____

E-MAIL_____
May we use your e-mail address for confirmations and other types of information? ☐ Yes   ☐ No

Order From Your Local Bookstore or Directly From
The Haworth Press, Inc.
10 Alice Street, Binghamton, New York 13904-1580 • USA
TELEPHONE: 1-800-HAWORTH (1-800-429-6784) / Outside US/Canada: (607) 722-5857
FAX: 1-800-895-0582 / Outside US/Canada: (607) 772-6362
E-mail: getinfo@haworthpressinc.com
PLEASE PHOTOCOPY THIS FORM FOR YOUR PERSONAL USE.

BOF96